Ready For 5th Grade

Math

A Summer Learning Activity Workbook For Students Entering Grade 5 [Ages 10-11]

Written and edited by Matthew Bowling

Printed in the United States of America

This paper meets the requirements of ANSI/NISO Z39.48-1992 (Permanence of Paper).

Mometrix offers volume discount pricing to institutions. For more information or a price quote, please contact our sales department at sales@mometrix.com or 888-248-1219.

Paperback
ISBN 13: 978-1-5167-3145-9
ISBN 10: 1-5167-3145-X

TABLE OF CONTENTS

Introduction

Congratulations on finishing 4th grade! That is a big accomplishment, and you should be proud of yourself.

As you get ready to move into **5th grade**, it is normal to feel a little nervous. This workbook is designed to help you feel prepared and confident as you take on new challenges. You've got what it takes to succeed.

This workbook is organized into chapters that review important 4th grade skills and introduce new concepts you'll see in 5th grade. Each chapter starts with a short lesson to refresh your memory or even introduce new things before you start the daily activities. You can always go back to the lesson page if you need extra help.

The workbook is made to be completed over 10 weeks, with each day taking no more than 15 minutes. That is just enough time to practice without feeling overwhelmed. As you work through the chapters, the activities will gradually become more challenging, helping you strengthen your thinking skills and prepare for the work you'll do in 5th grade.

Sometimes you may work with an adult, and other times you may work independently. Either way, you're building skills, confidence, and strong learning habits. Get ready, because 5th grade is going to be a great year!

Chapter 1: Whole Numbers

Rounding is used to change a number to a nearby number that is easier to work with. To round, look at the digit after the place you want to round to. For example, if you are asked to round to the nearest tens place, look at the ones place. If the digit is 0, 1, 2, 3, or 4, the number in the place you are rounding to stays the same. This is referred to as rounding down. For example, 342 rounded to the nearest tens place is 340. If the digit is 5, 6, 7, 8, or 9, the number in the place you are rounding to goes up to the next value. This is referred to as rounding up. For example, 768 rounded to the nearest tens place is 770.

Numbers can be broken apart and composed based on their **place value**. For example, the number 129,548 can be expressed in **expanded form** as 100,000 + 20,000 + 9,000 + 500 + 40 + 8. You should also be comfortable going from expanded form to standard form. For example, what number is represented by 30,000 + 400 + 80 + 2? Notice 3 groups of ten-thousands, 0 groups of thousands, 4 groups of hundreds, 8 groups of tens, and 2 groups of ones. This becomes 30,482.

Addition is when two or more sets of things are combined or joined together. This is also referred to as **composing**. The **sum** is the answer achieved after adding the sets. Similar to counting, unique symbols are used to indicate addition. The **plus sign** (+) represents addition, and the **equal sign** (=) is used to indicate the sum or answer.

To add multiple digit numbers together, start in the ones place and use basic addition rules. When the number equals ten or more, the first digit of the number carries over to the next place value. This is called **regrouping**. Look at the example below. When the numbers in the ones column add up to a number larger than 10, the first digit of that number carries over to the tens column.

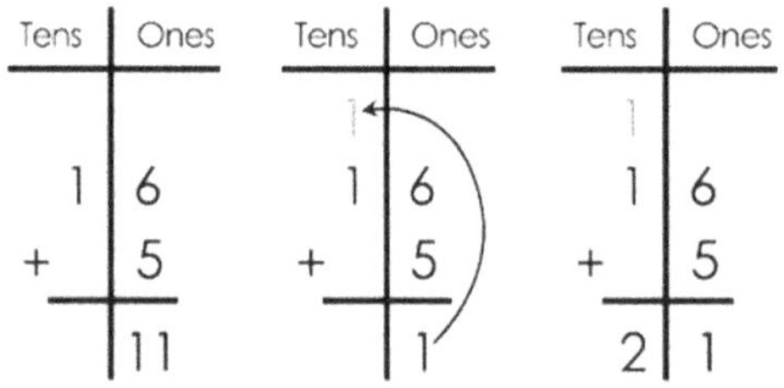

This same process is used for larger numbers as well.

Subtraction is when a set of things has a portion that is separated, removed, or taken away. This is also referred to as **decomposing**. The **difference** is the answer achieved after subtracting from the set. Similar to counting and addition, unique symbols are used to indicate subtraction. The **minus sign** (-) represents subtraction, and the **equal sign** (=) is used to indicate the difference or answer.

To subtract multiple digit numbers together, start in the ones place and use basic subtraction rules. If the top number in the column is smaller than the bottom, borrow a group of ten from the column to the left. This is called **borrowing**. Look at the example below.

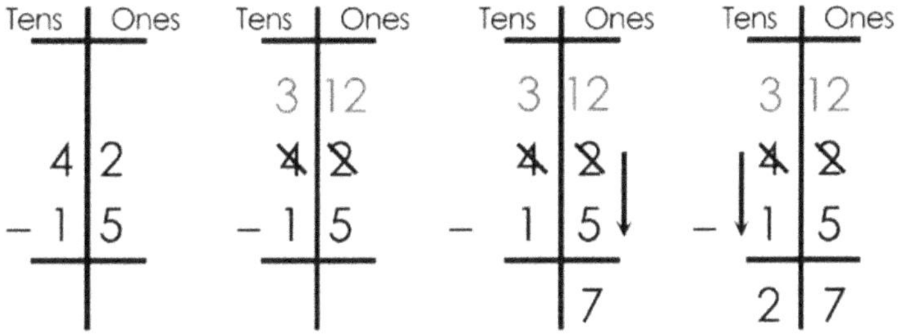

This same process is used for larger numbers as well.

Many problems can be solved using estimation. **Estimation** means calculating an **approximate** answer. Estimation can also be used as a strategy for checking the accuracy or reasonableness of an answer. For example, if the estimated amount of time it took you to run 3 miles was 30 minutes, but your calculation showed 300 minutes, you would know that there was an error in your math. Fluency, mental math, and strategies such as compatible numbers can make estimation a quick process. The goal is to use estimation as a quick reference point for the accuracy of an answer. For example, consider the following example:

Hank pays $231 for groceries. Betty pays $198 for groceries. About how much more did Hank pay than Betty?

Use estimation to solve. The question asks "about" how much more, which indicates that estimation will be sufficient. Consider rounding each number to the nearest ten. $231 becomes $230, and $198 becomes $200. The difference between $230 and $200 can be calculated mentally, $\$230 - \$200 = \$30$. Hank spent about $30 more than Betty.

Complete one worksheet each day in this chapter. Each day should take no longer than 15 minutes to complete.

Day 1

Place Value 1-100,000,000

The place value of a digit is determined by where it is in a number.

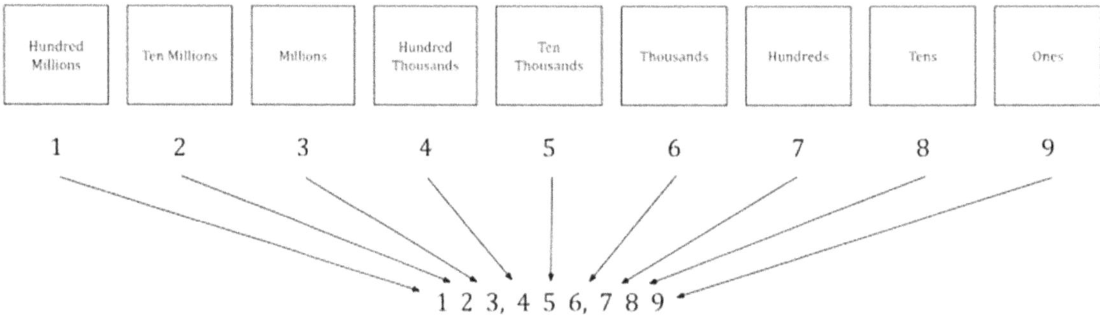

One Hundred Twenty-Three Million, Four Hundred Fifty-Six Thousand, Seven Hundred Eighty-Nine

Write these numbers correctly in the blanks below.

1. 748,422,719	=	Hundred Millions	Ten Millions	Millions	Hundred Thousands	Ten Thousands	Thousands	Hundreds	Tens	Ones
2. 329,608,114	=	Hundred Millions	Ten Millions	Millions	Hundred Thousands	Ten Thousands	Thousands	Hundreds	Tens	Ones
3. 124,375,277	=	Hundred Millions	Ten Millions	Millions	Hundred Thousands	Ten Thousands	Thousands	Hundreds	Tens	Ones
4. 741,588,379	=	Hundred Millions	Ten Millions	Millions	Hundred Thousands	Ten Thousands	Thousands	Hundreds	Tens	Ones
5. 504,267,332	=	Hundred Millions	Ten Millions	Millions	Hundred Thousands	Ten Thousands	Thousands	Hundreds	Tens	Ones

Day 2

Rounding up to 100,000

Round the following numbers to the nearest thousand.

1. 2,563
2. 9,198
3. 1,423
4. 7,712
5. 3,300
6. 4,219
7. 5,756
8. 8,154
9. 6,069
10. 1,995

Round the following numbers to the nearest ten-thousand.

11. 39,092
12. 19,917
13. 93,254
14. 56,055
15. 70,856
16. 77,150
17. 33,809
18. 35,451
19. 20,901
20. 48,599

Round the following numbers to the nearest hundred-thousand.

21. 274,333
22. 596,559
23. 221,324
24. 530,708
25. 189,365
26. 317,110
27. 882,658
28. 610,567
29. 789,381
30. 109,227

Day 3

Estimating

Solve the problems below using estimation.

1.

```
    8 5 8        9 0 0
+   3 9 1    +   4 0 0
---------    ---------
             1,  3 0 0
```

2.

```
    4 6 1
+   5 4 5    +
---------    ---------
```

3.

```
    1 2 9
+   6 0 7    +
---------    ---------
```

4.

```
    3 9 9
+   7 3 1    +
---------    ---------
```

5.

```
    9 4 4
-   4 9 9    -
---------    ---------
```

6.

```
    7 6 8
-   2 0 7    -
---------    ---------
```

7.

```
    6 3 0
-   4 5 2    -
---------    ---------
```

8.

```
    3 5 6
-   1 7 5    -
---------    ---------
```

Day 4

5-Digit Addition

Solve the problems below using regrouping.

1.	2.	3.	4.
54,691	90,861	44,388	50,233
10,847	77,392	35,512	46,397
+ 39,116	+ 20,691	+ 64,172	+ 92,846

5.	6.	7.	8.
58,732	19,570	90,691	35,450
33,020	27,694	51,072	82,300
+ 12,571	+ 56,630	+ 14,321	+ 23,124

9.	10.	11.	12.
60,805	66,349	15,000	99,785
84,997	70,244	39,289	35,562
+ 39,111	+ 11,188	+ 2,723	+ 48,078

Day 5

6-Digit Addition

Solve the problems below using regrouping.

1.

```
   102,220
   366,357
 + 127,475
 _________
```

2.

```
   557,365
   166,204
 + 254,129
 _________
```

3.

```
   325,299
   770,351
 +  64,658
 _________
```

4.

```
   129,195
   632,744
 + 335,163
 _________
```

5.

```
   641,001
   237,925
 + 210,137
 _________
```

6.

```
   752,112
   209,482
 + 322,165
 _________
```

7.

```
   480,129
   638,635
 + 218,001
 _________
```

8.

```
   500,051
   217,521
 + 127,937
 _________
```

9.

```
   637,817
   289,364
 + 123,994
 _________
```

Day 6

5-Digit Subtraction

Solve the problems below using borrowing.

1.	2.	3.	4.
25,911	40,127	68,956	42,207
- 23,125	- 15,089	- 47,580	- 5,699

5.	6.	7.	8.
30,113	92,253	57,115	21,659
- 9,357	- 75,449	- 36,608	- 17,102

9.	10.	11.	12.
81,678	47,827	35,157	50,011
- 44,169	- 9,018	- 16,658	- 23,956

Day 7

6-Digit Subtraction

Solve the problems below using regrouping.

1.

```
   627,343
 - 495,861
 ---------
```

2.

```
   854,115
 - 346,264
 ---------
```

3.

```
   506,925
 - 229,167
 ---------
```

4.

```
   129,239
 -  95,226
 ---------
```

5.

```
   958,062
 - 179,687
 ---------
```

6.

```
   297,343
 - 105,907
 ---------
```

7.

```
   961,343
 - 550,097
 ---------
```

8.

```
   639,447
 - 287,334
 ---------
```

9.

```
   339,081
 - 117,118
 ---------
```

Chapter 2: Multiplication

Multiplication is the process of counting equally sized groups. The multiplication sign (×) is read as times. For example, 5×7 is read as five times seven. This means 5 groups of 7 or $7 + 7 + 7 + 7 + 7$. The answer to a multiplication problem is called the **product**. Thus, the product of 5 groups of 7 is 35. The process of using addition to solve multiplication problems is known as **repeated addition**.

Factors are the numbers that are being multiplied together. For example, the factors in the multiplication problem $4 \times 3 = 12$ are four and three. Below shows a standard multiplication table with factors and products.

x	1	2	3	4	5	6	7	8	9	10
1	1	2	3	4	5	6	7	8	9	10
2	2	4	6	8	10	12	14	16	18	20
3	3	6	9	12	15	18	21	24	27	30
4	4	8	12	16	20	24	28	32	36	40
5	5	10	15	20	25	30	35	40	45	50
6	6	12	18	24	30	36	42	48	54	60
7	7	14	21	28	35	42	49	56	63	70
8	8	16	24	32	40	48	56	64	72	80
9	9	18	27	36	45	54	63	72	81	90
10	10	20	30	40	50	60	70	80	90	100

The **standard algorithm** is a commonly preferred strategy for numbers that are more than three digits. The standard algorithm, or "vertical multiplication," is less visual but more time-efficient. For example, consider the product $7{,}482 \times 8$. Write the longer number on the top and the shorter number on the bottom. Push all numbers to the right so that the place value is lined up (ones lined up with ones, tens lined up with tens, etc.).

$$\begin{array}{r} {}^{3\ 6\ 1} \\ 7{,}482 \\ \times \quad 8 \\ \hline 59{,}856 \end{array}$$

Distribute, or multiply, the 8 by each digit in the number 7,482. Work from right to left. 8×2, then 8×8, then 8×4, then 8×7. If a product has more than one digit, write the ones digit in the answer and carry the tens digit above the next number you are multiplying. For example, $2 \times 8 = 16$, so write 6 in the answer and 1 above 8 (the next number you will multiply by). Then, add the carried number to the next product. $8 \times 8 = 64$, then add 1 to get 65, and continue the multiplication process for the rest of the number. The product of 7,482 and 8 is 59,856.

To multiply two multi-digit numbers together, the process is very similar. Start by lining up the ones digit for both numbers and stack the two numbers vertically with the larger number on top. First multiply the digit in the ones place of the bottom number by all digits in the top number from right to left. Then, write a "place holder zero" beneath the initial product in the ones place. Repeat the process by multiplying the digit in the tens place of the bottom number by all the digits in the top number from right to left. Finally, add both products together to get the final product.

$$
\begin{array}{r}
1 \\
526 \\
\times\ 31 \\
\hline
{}^{1}_{1}526 \\
+\ 15780 \\
\hline
16{,}306
\end{array}
$$

Complete one worksheet each day in this chapter. Each day should take no longer than 15 minutes to complete.

Day 8

Multiplication

Solve the problems below.

1.
$$\begin{array}{r} 65 \\ \times \quad 4 \\ \hline 260 \end{array}$$

2.
$$\begin{array}{r} 49 \\ \times \quad 3 \\ \hline \end{array}$$

3.
$$\begin{array}{r} 12 \\ \times \quad 9 \\ \hline \end{array}$$

4.
$$\begin{array}{r} 92 \\ \times \quad 6 \\ \hline \end{array}$$

5.
$$\begin{array}{r} 57 \\ \times \quad 8 \\ \hline \end{array}$$

6.
$$\begin{array}{r} 49 \\ \times \quad 8 \\ \hline \end{array}$$

7.
$$\begin{array}{r} 29 \\ \times \quad 4 \\ \hline \end{array}$$

8.
$$\begin{array}{r} 32 \\ \times \quad 5 \\ \hline \end{array}$$

9.
$$\begin{array}{r} 68 \\ \times \quad 7 \\ \hline \end{array}$$

10.
$$\begin{array}{r} 83 \\ \times \quad 9 \\ \hline \end{array}$$

Day 9

Multiplying 2-Digit Numbers by 2-Digit Numbers

Solve the problems below.

1. $\begin{array}{r} 49 \\ \times\ 15 \\ \hline \end{array}$

2. $\begin{array}{r} 64 \\ \times\ 24 \\ \hline \end{array}$

3. $\begin{array}{r} 54 \\ \times\ 36 \\ \hline \end{array}$

4. $\begin{array}{r} 67 \\ \times\ 48 \\ \hline \end{array}$

5. $\begin{array}{r} 85 \\ \times\ 65 \\ \hline \end{array}$

6. $\begin{array}{r} 74 \\ \times\ 46 \\ \hline \end{array}$

7. $\begin{array}{r} 86 \\ \times\ 39 \\ \hline \end{array}$

8. $\begin{array}{r} 35 \\ \times\ 29 \\ \hline \end{array}$

9. $\begin{array}{r} 68 \\ \times\ 47 \\ \hline \end{array}$

10. $\begin{array}{r} 72 \\ \times\ 25 \\ \hline \end{array}$

11. $\begin{array}{r} 54 \\ \times\ 54 \\ \hline \end{array}$

12. $\begin{array}{r} 87 \\ \times\ 64 \\ \hline \end{array}$

13. $\begin{array}{r} 62 \\ \times\ 43 \\ \hline \end{array}$

14. $\begin{array}{r} 95 \\ \times\ 92 \\ \hline \end{array}$

15. $\begin{array}{r} 42 \\ \times\ 39 \\ \hline \end{array}$

Day 10

Multiplication Word Problems

Use multiplication to solve the problems below.

1. If Tommy plays 7 soccer games, and he blocks 6 shots in each game, how many shots will he block?

____

2. Larry loves photography and takes a lot of pictures. If he takes 32 pictures a day for 20 days, how many pictures will he take?

____

3. James is a great painter. He can paint 3 paintings a day. If he paints for 19 days, how many paintings will he have?

____

4. Brian eats pizza every day. If he eats 8 slices a day for 9 days, how many slices of pizza will he eat?

____

DAY 11

MULTIPLYING 3-DIGIT NUMBERS BY 2-DIGIT NUMBERS

Solve the problems below.

1.
$$\begin{array}{r} 321 \\ \times\ \ 23 \\ \hline 963 \\ +\ 6420 \\ \hline 7383 \end{array}$$

2.
$$\begin{array}{r} 259 \\ \times\ \ 19 \\ \hline \\ + \qquad \\ \hline \end{array}$$

3.
$$\begin{array}{r} 512 \\ \times\ \ 43 \\ \hline \\ + \qquad \\ \hline \end{array}$$

4.
$$\begin{array}{r} 459 \\ \times\ \ 36 \\ \hline \\ + \qquad \\ \hline \end{array}$$

5.
$$\begin{array}{r} 608 \\ \times\ \ 25 \\ \hline \\ + \qquad \\ \hline \end{array}$$

6.
$$\begin{array}{r} 641 \\ \times\ \ 39 \\ \hline \\ + \qquad \\ \hline \end{array}$$

7.
$$\begin{array}{r} 539 \\ \times\ \ 62 \\ \hline \\ + \qquad \\ \hline \end{array}$$

8.
$$\begin{array}{r} 235 \\ \times\ \ 24 \\ \hline \\ + \qquad \\ \hline \end{array}$$

9.
$$\begin{array}{r} 684 \\ \times\ \ 31 \\ \hline \\ + \qquad \\ \hline \end{array}$$

10.
$$\begin{array}{r} 843 \\ \times\ \ 64 \\ \hline \\ + \qquad \\ \hline \end{array}$$

11.
$$\begin{array}{r} 892 \\ \times\ \ 83 \\ \hline \\ + \qquad \\ \hline \end{array}$$

12.
$$\begin{array}{r} 399 \\ \times\ \ 95 \\ \hline \\ + \qquad \\ \hline \end{array}$$

13.
$$\begin{array}{r} 705 \\ \times\ \ 34 \\ \hline \\ + \qquad \\ \hline \end{array}$$

14.
$$\begin{array}{r} 633 \\ \times\ \ 79 \\ \hline \\ + \qquad \\ \hline \end{array}$$

15.
$$\begin{array}{r} 514 \\ \times\ \ 92 \\ \hline \\ + \qquad \\ \hline \end{array}$$

DAY 12

MULTIPLYING 4-DIGIT NUMBERS BY 2-DIGIT NUMBERS

Solve the problems below.

1.
$$\begin{array}{r} 1,491 \\ \times \quad 52 \\ \hline 2982 \\ +\ 74550 \\ \hline 77,532 \end{array}$$

2.
$$\begin{array}{r} 6,861 \\ \times \quad 15 \\ \hline \\ + \quad \\ \hline \end{array}$$

3.
$$\begin{array}{r} 4,532 \\ \times \quad 37 \\ \hline \\ + \quad \\ \hline \end{array}$$

4.
$$\begin{array}{r} 2,725 \\ \times \quad 93 \\ \hline \\ + \quad \\ \hline \end{array}$$

5.
$$\begin{array}{r} 3,116 \\ \times \quad 29 \\ \hline \\ + \quad \\ \hline \end{array}$$

6.
$$\begin{array}{r} 7,683 \\ \times \quad 49 \\ \hline \\ + \quad \\ \hline \end{array}$$

7.
$$\begin{array}{r} 4,492 \\ \times \quad 66 \\ \hline \\ + \quad \\ \hline \end{array}$$

8.
$$\begin{array}{r} 2,374 \\ \times \quad 97 \\ \hline \\ + \quad \\ \hline \end{array}$$

9.
$$\begin{array}{r} 5,051 \\ \times \quad 31 \\ \hline \\ + \quad \\ \hline \end{array}$$

10.
$$\begin{array}{r} 8,790 \\ \times \quad 29 \\ \hline \\ + \quad \\ \hline \end{array}$$

11.
$$\begin{array}{r} 2,778 \\ \times \quad 86 \\ \hline \\ + \quad \\ \hline \end{array}$$

12.
$$\begin{array}{r} 6,627 \\ \times \quad 61 \\ \hline \\ + \quad \\ \hline \end{array}$$

DAY 13

MULTIPLYING 3-DIGIT NUMBERS BY 3-DIGIT NUMBERS

Solve the problems below.

1.

$$\begin{array}{r} 432 \\ \times\ 365 \\ \hline 2160 \\ 25920 \\ +\ 129600 \\ \hline 157{,}680 \end{array}$$

2.

$$\begin{array}{r} 359 \\ \times\ 619 \\ \hline \\ + \\ \hline \end{array}$$

3.

$$\begin{array}{r} 482 \\ \times\ 165 \\ \hline \\ + \\ \hline \end{array}$$

4.

$$\begin{array}{r} 742 \\ \times\ 363 \\ \hline \\ + \\ \hline \end{array}$$

5.

$$\begin{array}{r} 476 \\ \times\ 134 \\ \hline \\ + \\ \hline \end{array}$$

6.

$$\begin{array}{r} 895 \\ \times\ 135 \\ \hline \\ + \\ \hline \end{array}$$

7.

$$\begin{array}{r} 922 \\ \times\ 617 \\ \hline \\ + \\ \hline \end{array}$$

8.

$$\begin{array}{r} 384 \\ \times\ 475 \\ \hline \\ + \\ \hline \end{array}$$

9.

$$\begin{array}{r} 524 \\ \times\ 277 \\ \hline \\ + \\ \hline \end{array}$$

Day 14

Multiplying 4-Digit Numbers by 3-Digit Numbers

Solve the problems below.

1.

$$\begin{array}{r} 1{,}291 \\ \times\ 324 \\ \hline 5164 \\ 25820 \\ +\ 387300 \\ \hline 418{,}284 \end{array}$$

2.

$$\begin{array}{r} 2{,}177 \\ \times\ 115 \\ \hline \\ + \\ \hline \end{array}$$

3.

$$\begin{array}{r} 4{,}264 \\ \times\ 137 \\ \hline \\ + \\ \hline \end{array}$$

4.

$$\begin{array}{r} 3{,}387 \\ \times\ 244 \\ \hline \\ + \\ \hline \end{array}$$

5.

$$\begin{array}{r} 5{,}189 \\ \times\ 320 \\ \hline \\ + \\ \hline \end{array}$$

6.

$$\begin{array}{r} 3{,}654 \\ \times\ 439 \\ \hline \\ + \\ \hline \end{array}$$

7.

$$\begin{array}{r} 5{,}307 \\ \times\ 148 \\ \hline \\ + \\ \hline \end{array}$$

8.

$$\begin{array}{r} 2{,}273 \\ \times\ 259 \\ \hline \\ + \\ \hline \end{array}$$

9.

$$\begin{array}{r} 4{,}892 \\ \times\ 341 \\ \hline \\ + \\ \hline \end{array}$$

Chapter 3: Division

Division is a way of determining how many equal groups can be made out of the given amount. For example, 28 bowlers can be divided evenly into 7 groups of 4 people.

This kind of problem can be solved using a variety of strategies. A sketch or visual can be made to evenly partition the bowlers into groups. Multiplication can also be used to determine the answer to the division problem or **quotient**. Similar to addition and subtraction, multiplication and division are **inverse operations**. For example, $72 \div 9$ can be solved using the multiplication facts for 9. $72 \div 9 = ?$ Ask the question, "9 groups of what equals 72?" $9 \times 6 = 54 \ldots 9 \times 7 = 63 \ldots 9 \times 8 = 72$. The missing number is 8. This means $72 \div 9 = 8$. This strategy is quite helpful because once the multiplication facts are automatic, division facts become automatic as well.

The **division sign** ($\div$ or $\overline{)}$) means "divided by." The **dividend** is the number that is divided or broken up into parts. The **divisor** is the number that tells how many parts the dividend will be broken up into.

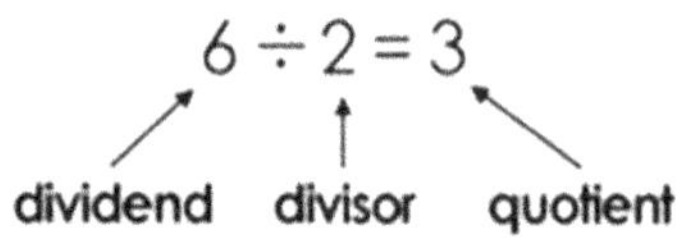

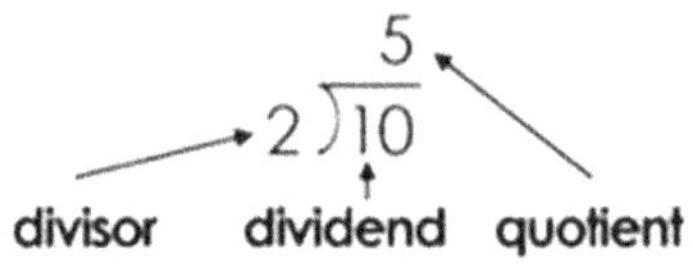

The problem to the left is read as "six divided by two is equal to three," and the problem to the right is read as "ten divided by two is equal to five."

Sometimes groups of objects or numbers cannot be divided into equal groups, and the number left over in a division problem is called the **remainder**. The remainder must be smaller than the divisor. For example, if 13 is divided into 5 equal groups, there will be 2 equal groups with 3 left over.

$$\begin{array}{r} 2\text{r}3 \\ 5\overline{)13} \\ -10 \\ \hline 3 \end{array}$$

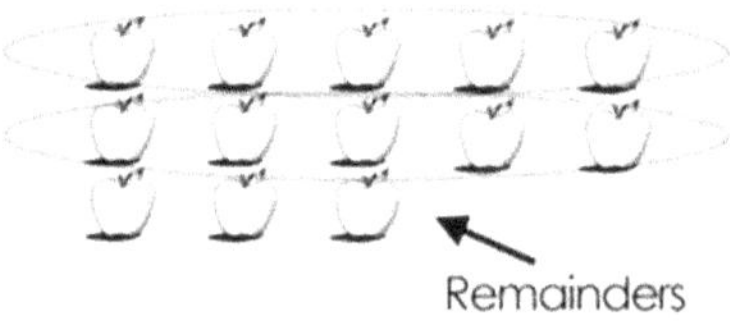

The **standard algorithm** for division, also known as **long division**, is a great option for division problems involving large numbers. To complete long division, the standard process is to estimate, divide, multiply, subtract, compare, bring down, and repeat as necessary. This means start by looking at the first digit in the dividend, and estimating how many times the divisor can evenly fit into the number.

For example, when dividing 3,190 by 58, place 3,190 under the symbol and 58 to the left of the symbol.

$3{,}190 \div 58$ becomes:

$$58\overline{)3190}$$

Step 1: Figure out how many times the divisor can fit into the first digit of the dividend without going over it. In this case, 58 cannot go into 3 because 3 is too small. Include the next digit. Again, 58 cannot go into 31 because 31 is too small. Include the next digit. 58 can go into 319 because 319 is larger than 58. Use estimation to determine how many times 58 goes into 319. Round 58 to 60. Round 319 to 300. 60 goes into 300 5 times. This estimation can be used to determine how many times 58 goes into 319. Write 5 above the 9. Make sure the 5 is above the hundreds digit because this is the farthest place value considered in the estimation.

$$\begin{array}{r} 5 \\ 58\overline{)3190} \end{array}$$

Step 2: Multiply 5×58. The product of 5×58 is 290. Write 290 below the dividend.

$$\begin{array}{r} 5 \\ 58\overline{)3190} \\ 290 \end{array}$$

Step 3: Subtract $319 - 290 = 29$.

$$\begin{array}{r} 5 \\ 58\overline{)3190} \\ \underline{-290} \\ 29 \end{array}$$

Step 4: Bring down the next digit and repeat the process.

$$\begin{array}{r} 55 \\ 58\overline{)3190} \\ \underline{-290} \\ 290 \\ \underline{-290} \\ 0 \end{array}$$

The last value is 0. This indicates that there is no remainder and that $3{,}190 \div 58 = 55$.

Complete one worksheet each day in this chapter. Each day should take no longer than 15 minutes to complete.

Day 15

Division with a 2-Digit Dividend

Find the quotients. Some may or may not have remainders.

1. $5\overline{)67}$

2. $8\overline{)99}$

3. $3\overline{)53}$

4. $9\overline{)92}$

5. $2\overline{)78}$

6. $7\overline{)80}$

7. $5\overline{)73}$

8. $3\overline{)64}$

9. $4\overline{)61}$

10. $2\overline{)83}$

11. $9\overline{)97}$

12. $6\overline{)77}$

Day 16

Division Word Problems

Use division to solve the problems below.

1. Jerry has 24 boxes to move. He can only carry 3 boxes at a time. How many trips will he have to make to move all the boxes?

____

2. Sandy has 15 chores to do around the house this week. If she does 3 chores a day, how many days will it take her to do all her chores?

____

3. Mitchell has 42 apples. He wants to divide them into 6 groups. How many apples will be in each group?

____

4. Andy has 63 cookies in his jar that he wants to share with his friends. If he has 7 friends, how many cookies will each friend get?

____

Day 17

Dividing 3-Digit Dividends by a 1-Digit Divisor

Find the quotients. Some may or may not have remainders.

1.

```
    1 5 3 r 2
3 ) 4 6 1
  - 3
    1 6
  - 1 5
      1 1
    -   9
        2
```

2. 4) 1 2 8

3. 2) 3 9 2

4. 7) 9 7 5

5. 9) 8 1 5

6. 6) 8 2 9

7. 3) 5 6 4

8. 2) 2 9 7

9. 3) 6 2 2

10. 4) 9 3 0

11. 5) 5 5 6

12. 7) 9 8 6

13. 2) 8 4 6

14. 3) 1 3 4

15. 2) 4 0 8

Day 18

Dividing 4-Digit Dividends by a 1-Digit Divisor

Find the quotients. Some may or may not have remainders.

1. $5\overline{)6,207}$

2. $8\overline{)4,557}$

3. $2\overline{)3,054}$

4. $9\overline{)5,671}$

5. $7\overline{)9,278}$

6. $4\overline{)8,715}$

7. $3\overline{)1,896}$

8. $6\overline{)5,809}$

9. $5\overline{)2,254}$

10. $2\overline{)3,735}$

11. $6\overline{)6,854}$

12. $9\overline{)8,994}$

DAY 19

DIVISION WITH 2-DIGIT DIVISORS

Find the quotients. Some may or may not have remainders.

1. $31\overline{)659}$

2. $46\overline{)884}$

3. $21\overline{)429}$

4. $19\overline{)708}$

5. $38\overline{)594}$

6. $75\overline{)999}$

7. $53\overline{)864}$

8. $60\overline{)729}$

9. $91\overline{)963}$

10. $43\overline{)836}$

Day 20

Dividing 3-Digit Dividends by a 2-Digit Divisor

Find the quotients. Some may or may not have remainders.

1. $12\overline{)269}$
2. $30\overline{)738}$
3. $22\overline{)449}$
4. $41\overline{)209}$
5. $20\overline{)861}$
6. $12\overline{)271}$
7. $45\overline{)994}$
8. $33\overline{)735}$
9. $35\overline{)755}$
10. $11\overline{)195}$
11. $29\overline{)312}$
12. $63\overline{)656}$

Day 21

Dividing 4-Digit Dividends by a 2-Digit Divisor

Find the quotients. Some may or may not have remainders.

1. $13\overline{)2{,}691}$

2. $24\overline{)3{,}558}$

3. $65\overline{)8{,}945}$

4. $54\overline{)6{,}021}$

5. $41\overline{)9{,}278}$

6. $23\overline{)5{,}968}$

7. $36\overline{)7{,}682}$

8. $17\overline{)4{,}950}$

9. $44\overline{)8{,}962}$

10. $16\overline{)1{,}891}$

11. $21\overline{)3{,}398}$

12. $32\overline{)6{,}570}$

Chapter 4: Fractions

Fractions are numbers that make up part of a whole. The shaded parts represent the top of the fractions, which are called **numerators**. The bottom numbers, **denominators**, represent the total number of equal pieces each shape is broken up into.

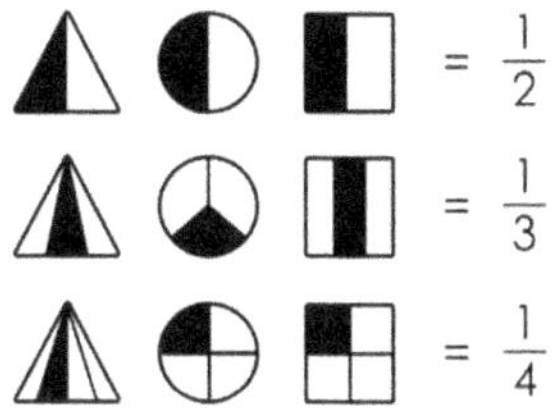

Mixed numbers are values that are written using a whole number and a fraction. For example, $2\frac{1}{3}$ is a mixed number that consists of 7 groups of $\frac{1}{3}$. **Improper fractions** are fractions that have a numerator that is greater than the denominator. $\frac{4}{3}$ is an improper fraction that consists of 4 groups of $\frac{1}{3}$.

To convert a mixed number to an improper fraction, begin by multiplying the whole number by the denominator of the fraction. Next, add the product to the numerator of the fraction. This value becomes the numerator for the improper fraction. The denominator used in the mixed number will be the denominator used in the improper fraction. For example, $2\frac{3}{4} = \frac{11}{4}$.

To convert an improper fraction to a mixed number, divide the numerator by the denominator: the quotient is the integer part of the mixed number, and the remainder is the numerator. For example, $27 \div 8 = 3$ with a remainder of 3, so $\frac{27}{8} = 3\frac{3}{8}$.

Fractions are considered **equivalent** if they represent the same amount. You can make equivalent fractions through multiplication or division. For example, $\frac{6}{12} = \frac{18}{36}$. To achieve this, both the numerator and denominator of the starting fraction are multiplied by 3. Likewise, $\frac{18}{36} = \frac{6}{12}$. To achieve this, both the numerator and denominator of the starting fraction are divided by 3.

Fractions with the same denominator can be added or subtracted by adding or subtracting the numerators and keeping the denominators the same. Fractions with different denominators must first be made into an equivalent fraction so that the denominators are the same for each fraction involved in the addition or subtraction problem.

Simplifying fractions means reducing a fraction to the lowest possible equivalent term. To do this, find the greatest common factor (GCF) of the numerator and the denominator. Then divide both the numerator and denominator by that factor.

Complete one worksheet each day in this chapter. Each day should take no longer than 15 minutes to complete.

Day 22

Adding and Subtracting Fractions with Common Denominators

Add or subtract the fractions below.

1.

$$\frac{6}{15}+\frac{5}{15}+\frac{2}{15}= \qquad \frac{13}{15}$$

2.

$$\frac{12}{37}+\frac{7}{37}+\frac{4}{37}=$$

3.

$$\frac{13}{80}+\frac{9}{80}+\frac{27}{80}=$$

4.

$$\frac{8}{109}+\frac{77}{109}+\frac{15}{109}=$$

5.

$$\frac{17}{64}+\frac{3}{64}+\frac{10}{64}=$$

6.

$$\frac{3}{21}+\frac{7}{21}+\frac{4}{21}=$$

7.

$$\frac{608}{759}-\frac{325}{759}=$$

8.

$$\frac{45}{56}-\frac{19}{56}=$$

9.

$$\frac{178}{207}-\frac{99}{207}=$$

10.

$$\frac{30}{49}-\frac{17}{49}=$$

11.

$$\frac{246}{277}-\frac{79}{277}=$$

12.

$$\frac{439}{534}-\frac{201}{534}=$$

Day 23

Fraction Word Problems

Write the correct fractions for each word problem below.

1. Daryl is working on his garden. He has planted 28 plants so far. He has 13 tomato plants, 6 eggplants, and 9 potato plants. Write the fraction that represents each type of plant out of the total amount of plants in his garden.

Tomatoes: $\frac{13}{28}$ ______ Eggplants: ______ Potatoes: ______

2. Mom's favorite hobby is bird watching. Today she saw 15 different birds. She saw 3 Blue Jays, 9 Sparrows, 6 Ducks, and 1 Eagle. Write the fraction that represents each type of bird out of the total amount of birds she saw.

Blue Jays: ______ Sparrows: ______ Ducks: ______ Eagles: ______

3. Johnny went fishing today and caught 17 fish. He caught 2 bass, 6 trout, 5 guppies, and 4 goldfish. Write the fraction that represents each type of fish out of the total amount of fish he caught.

Bass: ______ Trout: ______ Guppies: ______ Goldfish: ______

4. Dad is barbequing for the party this afternoon. He has 76 items to cook. He is going to cook 15 hamburgers, 22 chicken wings, and 39 sausages. Write the fraction that represents each type of food item he is going to cook out of the total amount of item.

Hamburgers: ______ Chicken Wings: ______ Sausages: ______

DAY 24

EQUIVALENT FRACTIONS

Shade in the fractions on the right to equal the fractions on the left.

1.

$\frac{3}{8}$ 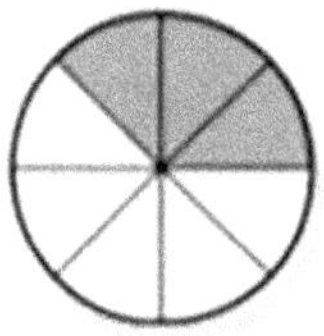 = $\frac{6}{16}$ 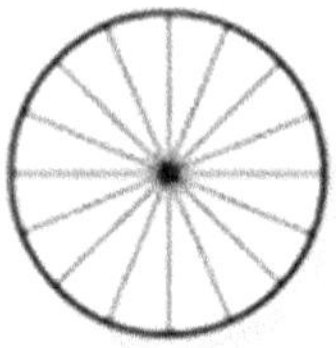

2.

$\frac{5}{8}$ 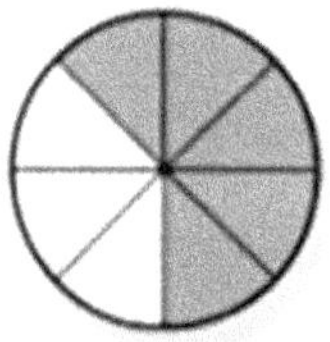 = $\frac{10}{16}$ 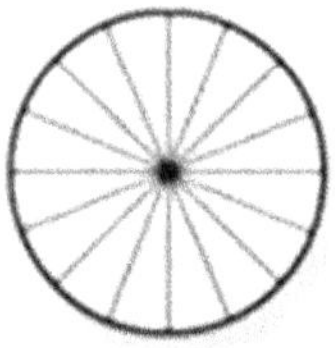

Finish writing the problems out to reduce the fractions to their simplest form.

3. $\frac{18}{27} \div \frac{9}{9} = \frac{2}{3}$

4. $\frac{4}{12} \div \underline{\qquad} = \underline{\qquad}$

5. $\frac{20}{35} \div \underline{\qquad} = \underline{\qquad}$

6. $\frac{9}{24} \div \underline{\qquad} = \underline{\qquad}$

7. $\frac{15}{21} \div \underline{\qquad} = \underline{\qquad}$

8. $\frac{16}{40} \div \underline{\qquad} = \underline{\qquad}$

Solve each problem by making an equivalent fraction and then finding the sum or difference.

9.

$$\frac{6}{10}+\frac{1}{5} = \mathbf{\frac{3}{5}}+\frac{1}{5} = \mathbf{\frac{4}{5}}$$

10.

$$\frac{9}{21}+\frac{2}{7} = \qquad +\frac{2}{7} =$$

11.

$$\frac{15}{25}-\frac{1}{5} = \qquad -\frac{1}{5} =$$

12.

$$\frac{25}{35}-\frac{3}{7} = \qquad -\frac{3}{7} =$$

Day 25

Adding and Subtracting Fractions with Different Denominators Using Division

Solve each problem by finding the common denominator using division, creating an equivalent fraction, and then finding the sum.

1. $\frac{15}{21} + \frac{2}{7} = \frac{5}{7} + \frac{2}{7} = \frac{7}{7}$

2. $\frac{1}{5} + \frac{8}{20} = \frac{\_}{\_} + \frac{\_}{\_} = \frac{\_}{\_}$

3. $\frac{21}{30} + \frac{2}{10} = \frac{\_}{\_} + \frac{\_}{\_} = \frac{\_}{\_}$

4. $\frac{10}{12} + \frac{6}{36} = \frac{\_}{\_} + \frac{\_}{\_} = \frac{\_}{\_}$

5. $\frac{2}{9} + \frac{18}{54} = \frac{\_}{\_} + \frac{\_}{\_} = \frac{\_}{\_}$

6. $\frac{25}{35} + \frac{2}{7} = \frac{\_}{\_} + \frac{\_}{\_} = \frac{\_}{\_}$

7. $\frac{81}{81} - \frac{6}{9} = \frac{\_}{\_} - \frac{\_}{\_} = \frac{\_}{\_}$

8. $\frac{8}{9} - \frac{12}{36} = \frac{\_}{\_} - \frac{\_}{\_} = \frac{\_}{\_}$

9. $\frac{6}{7} - \frac{21}{49} = \frac{\_}{\_} - \frac{\_}{\_} = \frac{\_}{\_}$

10. $\frac{8}{9} - \frac{12}{54} = \frac{\_}{\_} - \frac{\_}{\_} = \frac{\_}{\_}$

11. $\frac{5}{6} - \frac{16}{48} = \frac{\_}{\_} - \frac{\_}{\_} = \frac{\_}{\_}$

12. $\frac{32}{36} - \frac{5}{9} = \frac{\_}{\_} - \frac{\_}{\_} = \frac{\_}{\_}$

Day 26

Finding Common Denominators Using Multiplication

Fill in the blanks to make the fractions equivalent.

1. $\frac{3}{5} = \frac{\_\_}{20}$

2. $\frac{2}{4} = \frac{6}{\_\_}$

3. $\frac{1}{2} = \frac{\_\_}{20}$

4. $\frac{6}{8} = \frac{\_\_}{30}$

5. $\frac{1}{7} = \frac{\_\_}{35}$

6. $\frac{2}{6} = \frac{30}{\_\_}$

7. $\frac{4}{5} = \frac{\_\_}{100}$

8. $\frac{2}{4} = \frac{\_\_}{32}$

9. $\frac{4}{12} = \frac{\_\_}{84}$

10. $\frac{5}{10} = \frac{45}{\_\_}$

11. $\frac{3}{7} = \frac{6}{\_\_}$

12. $\frac{3}{5} = \frac{33}{\_\_}$

13. $\frac{9}{16} = \frac{72}{\_\_}$

14. $\frac{1}{3} = \frac{\_\_}{45}$

15. $\frac{15}{25} = \frac{30}{\_\_}$

16. $\frac{16}{18} = \frac{\_\_}{90}$

17. $\frac{3}{16} = \frac{\_\_}{64}$

18. $\frac{3}{5} = \frac{\_\_}{25}$

19. $\frac{4}{6} = \frac{4}{\_\_}$

20. $\frac{2}{8} = \frac{4}{\_\_}$

DAY 27

ADDING AND SUBTRACTING FRACTIONS WITH DIFFERENT DENOMINATORS USING MULTIPLICATION

Solve each problem by finding the common denominator using multiplication, creating an equivalent fraction, and then finding the sum.

1. $\frac{2}{4} + \frac{1}{8} = \frac{4}{8} + \frac{1}{8} = \frac{5}{8}$

2. $\frac{4}{12} + \frac{3}{6} = \frac{\ }{\ } + \frac{\ }{\ } = \frac{\ }{\ }$

3. $\frac{2}{7} + \frac{5}{21} = \frac{\ }{\ } + \frac{\ }{\ } = \frac{\ }{\ }$

4. $\frac{4}{28} + \frac{3}{4} = \frac{\ }{\ } + \frac{\ }{\ } = \frac{\ }{\ }$

5. $\frac{3}{5} + \frac{12}{50} = \frac{\ }{\ } + \frac{\ }{\ } = \frac{\ }{\ }$

6. $\frac{6}{24} + \frac{5}{8} = \frac{\ }{\ } + \frac{\ }{\ } = \frac{\ }{\ }$

7. $\frac{36}{49} - \frac{3}{7} = \frac{\ }{\ } - \frac{\ }{\ } = \frac{\ }{\ }$

8. $\frac{5}{6} - \frac{29}{54} = \frac{\ }{\ } - \frac{\ }{\ } = \frac{\ }{\ }$

9. $\frac{63}{81} - \frac{6}{9} = \frac{\ }{\ } - \frac{\ }{\ } = \frac{\ }{\ }$

10. $\frac{54}{63} - \frac{5}{9} = \frac{\ }{\ } - \frac{\ }{\ } = \frac{\ }{\ }$

11. $\frac{27}{48} - \frac{4}{8} = \frac{\ }{\ } - \frac{\ }{\ } = \frac{\ }{\ }$

12. $\frac{35}{50} - \frac{50}{100} = \frac{\ }{\ } - \frac{\ }{\ } = \frac{\ }{\ }$

DAY 28

MIXED NUMBERS WITH COMMON DENOMINATORS

Add or subtract the mixed numbers below.

1. $3\frac{6}{12} + 2\frac{4}{12} = 5\frac{10}{12}$

2. $5\frac{3}{9} + 1\frac{1}{9} = \_\frac{}{}$

3. $4\frac{2}{5} + 3\frac{2}{5} = \_\frac{}{}$

4. $1\frac{8}{23} + 1\frac{12}{23} = \_\frac{}{}$

5. $9\frac{2}{8} + 4\frac{4}{8} = \_\frac{}{}$

6. $6\frac{3}{15} + 7\frac{7}{15} = \_\frac{}{}$

7. $8\frac{21}{39} + 5\frac{9}{39} = \_\frac{}{}$

8. $3\frac{4}{10} + 9\frac{5}{10} = \_\frac{}{}$

9. $6\frac{12}{19} - 2\frac{7}{19} = \_\frac{}{}$

10. $8\frac{7}{8} - 5\frac{2}{8} = \_\frac{}{}$

11. $10\frac{14}{16} - 7\frac{8}{16} = \_\frac{}{}$

12. $14\frac{5}{6} - 9\frac{1}{6} = \_\frac{}{}$

13. $24\frac{21}{25} - 19\frac{17}{25} = \_\frac{}{}$

14. $36\frac{9}{10} - 25\frac{8}{10} = \_\frac{}{}$

15. $46\frac{39}{44} - 38\frac{12}{44} = \_\frac{}{}$

Day 29

Reducing Mixed Numbers with Common Denominators

Add or subtract the mixed numbers below. Then reduce the mixed number to its simplest form.

1. $4\frac{3}{10} + 1\frac{2}{10}$

   $5\frac{5}{10} = 5\frac{1}{2}$

2. $4\frac{8}{12} + 5\frac{1}{12}$

   — = —

3. $3\frac{2}{18} + 6\frac{4}{18}$

   — = —

4. $3\frac{1}{8} + 5\frac{3}{8}$

   — = —

5. $7\frac{7}{24} + 5\frac{5}{24}$

   — = —

6. $9\frac{10}{36} + 6\frac{8}{36}$

   — = —

7. $12\frac{10}{16} - 9\frac{2}{16}$

   — = —

8. $17\frac{5}{12} - 5\frac{1}{12}$

   — = —

9. $11\frac{21}{28} - 3\frac{7}{28}$

   — = —

10. $29\frac{9}{9} - 17\frac{3}{9}$

    — = —

11. $48\frac{37}{42} - 9\frac{9}{42}$

    — = —

12. $72\frac{25}{63} - 54\frac{7}{63}$

    — = —

Day 30

Converting Mixed Numbers and Improper Fractions

Convert each mixed number to an improper fraction.

1. $3\frac{4}{3} = \quad \frac{13}{3}$

2. $8\frac{6}{4} =$

3. $2\frac{9}{2} =$

4. $6\frac{2}{5} =$

5. $9\frac{5}{10} =$

6. $4\frac{1}{5} =$

7. $3\frac{4}{7} =$

8. $4\frac{2}{3} =$

Convert these improper fractions to whole numbers.

9. $\frac{35}{5} = 7$

10. $\frac{12}{3} =$

11. $\frac{24}{8} =$

12. $\frac{27}{3} =$

13. $\frac{48}{6} =$

14. $\frac{93}{3} =$

15. $\frac{25}{5} =$

16. $\frac{84}{2} =$

Convert these improper fractions to mixed numbers.

17. $\frac{15}{6} = 2\frac{3}{6}$

18. $\frac{10}{4} =$

19. $\frac{7}{2} =$

20. $\frac{19}{8} =$

21. $\frac{23}{4} =$

22. $\frac{26}{3} =$

23. $\frac{50}{9} =$

24. $\frac{46}{8} =$

Chapter 5: Decimals

The important thing to remember when adding and subtracting decimal numbers is to **line up** the **decimal points**. Lining up the decimal points ensures that similar place values are being combined. For example, ones combine with ones, tens combine with tens.

Consider the following: A pizza costs $27.50 and a drink costs $4.65. The total cost of the meal can be calculated by finding the sum. Stack the numbers vertically and line up the decimal points. With addition, it does not matter which number is on top or bottom. Regroup if necessary and bring the decimal straight down into the answer line. This method is known as the **standard algorithm**.

```
   11
   27. 50
 +  4. 65
 --------
 $32. 15
```

The standard algorithm is also useful for subtraction scenarios. For example, if a piece of ribbon measures 132.1 centimeters, and 27.76 centimeters are cut off and removed, how many centimeters of ribbon are left?

Stack the values vertically and line up the decimals. Make sure the larger number, 132.1, is on the top. Sometimes it is helpful to fill in any empty spaces with a zero. Regroup if necessary, and again bring the decimal straight down into the answer line.

```
        11   10
      2 1̸    0̸10
    1 3̸ 2̸ . 1̸ 0̸
 -    2 7 . 7 6
 ---------------
    1 0 4 . 3 4
```

There are 104.34 centimeters of ribbon left.

Money is expressed in **dollars** and **cents**. Cents are represented as **tenths** and **hundredths**. For example, $4.92 represents 4 whole dollars, 9 tenths of a dollar, and 2 hundredths of a dollar. These same rules are used to add or subtract money.

When multiplying decimals, you use the same strategy as multiplying whole numbers. The difference is that you have digits to the right of the decimal. When you are setting up a multiplication problem, the number with more digits will be on the top, and the number with fewer digits will be on the bottom. Set up this problem traditionally, with no decimal places. The number 23,142 has more digits than 45, so 23,142 will be at the top of the problem, and 45 will be at the bottom.

$$
\begin{array}{r}
\scriptstyle 1\ \ \ 1\ \ \ \ \\
\scriptstyle \not{1}\ \ \not{2}\,\not{1}\ \ \ \ \\
23{,}142 \\
\times \qquad 45 \\
\hline
115{,}710 \\
+\ \ 925{,}680 \\
\hline
1{,}041{,}390
\end{array}
$$

When using this strategy, add the decimal at the end. To figure out how many decimal places the answer will have, you count the number of digits behind the decimal point in each factor to find the decimal in the product. In this example, there are 3 decimal places in the first number (23.142) and 3 in the second number (0.045), so the answer will have 6 decimal places. Once the product is found, you would count back 6 digits and place the decimal in the appropriate place.

1041390.

So, 23.142 times 0.045 is 1.04139, which can be rounded to the nearest thousandth to get 1.041.

Complete one worksheet each day in this chapter. Each day should take no longer than 15 minutes to complete.

DAY 31

ARRANGING AND ADDING DECIMALS

Line up the decimal values and add them together.

1. $10.54 + 8.01 + 0.89$

2. $6.32 + 9.51 + 1.99$

3. $0.99 + 5.15 + 12.47$

4. $4.21 + 3.57 + 15.05$

5. $11.97 + 246.1 + 6.5$

6. $0.23 + 13.31 + 72.08$

7. $61.59 + 31.64 + 0.08$

8. $84 + 16.99 + 14.89$

9. $8.50 + 7.27 + 985$

Day 32

Decimal Addition Word Problems

Solve the problems below.

1. Jeremy is a great basketball player. In season one he averaged 9.3 points with 6.2 rebounds per game. In season two he averaged 24.9 points and 3.6 rebounds per game. In season three he averaged 41.2 points with 6.4 rebounds per game. What was the total of his points and the total number of rebound averages for all three seasons?

2. Randy eats way too much ice cream. He ate 1.5 ounces on Monday, 2.3 ounces on Tuesday, and 3.6 ounces on Wednesday. On Thursday he ate 6.9 ounces. Then on Friday he ate 4.2 ounces of ice cream. How much ice cream did Randy eat in total this week?

3. Cynthia has a parrot named Hank. Hank gets his exercise everyday by flying. On Monday he flew 5.1 miles. On Tuesday he flew 6.85 miles. On Wednesday he flew 2.97 miles. Then on Thursday he flew 4.63 miles. How many miles has Hank flown this week?

4. Dad went fishing today. The first fish he caught weighed 12.39 pounds. The second fish he caught weighed 16.05 pounds. The third fish he caught weighed 2.76 pounds. The last fish he caught weighed 9.14 pounds. What is the total weight of all the fish that Dad caught today?

Day 33

Adding Decimals

Add the values to find the totals.

1.

```
  123.47
   34.10
+   8.36
--------
```

2.

```
  106.05
   23.21
+   1.36
--------
```

3.

```
  253.75
   73.36
+  10.15
--------
```

4.

```
  387.12
   49.70
+   6.65
--------
```

5.

```
  513.08
  309.12
+  76.59
--------
```

6.

```
  480.21
  759.96
+  88.07
--------
```

7.

```
  624.71
  267.35
+  20.99
--------
```

8.

```
  945.37
  651.83
+  79.41
--------
```

9.

```
  11.892
  20.624
+ 57.393
+  8.102
--------
```

10.

```
   0.937
  13.419
  23.757
+  9.144
--------
```

11.

```
  12.587
   6.794
  90.725
+ 24.431
--------
```

12.

```
  67.599
  11.180
   9.413
+ 21.924
--------
```

Day 34

Arranging and Subtracting Decimals

Line up the decimal values and subtract them.

1. $72.53 - 22.56$

2. $99.58 - 71.90$

3. $33.29 - 19.4$

4. $40.06 - 24.1$

5. $59.19 - 58.02$

6. $82.33 - 51.73$

7. $29.23 - 11.06$

8. $95.19 - 48.44$

9. $80.46 - 62.9$

DAY 35

SUBTRACTING DECIMALS

Subtract the values to find the difference.

1.
```
  231.57
- 182.24
--------
```

2.
```
  432.66
- 106.27
--------
```

3.
```
  379.20
- 273.94
--------
```

4.
```
  628.12
- 504.39
--------
```

5.
```
  611.32
- 369.48
--------
```

6.
```
  461.30
- 119.82
--------
```

7.
```
  64.982
- 20.191
--------
```

8.
```
  32.100
- 12.086
--------
```

9.
```
  52.604
- 39.335
--------
```

10.
```
  11.325
-  5.425
--------
```

11.
```
  67.835
- 46.105
--------
```

12.
```
  90.488
- 53.914
--------
```

Day 36

Decimal Subtraction Word Problems

Solve the problems below.

1. Mike needs to swim 6 miles a week to get strong enough to make the swim team. On Monday he swam 1.34 miles. On Tuesday he swam 2.49 miles. On Wednesday he swam 0.82 miles. How much further must he swim to reach his goal of 6 miles?

____

2. Jenny wants to lift 100 pounds of weight each day. First, she lifted 10.57 pounds, and then she lifted 29.96 pounds. Next, she lifted 32.65 pounds. How much more weight does she need to lift to reach her goal?

____

3. Newt needs a total of 300 points on his school projects to go on the field trip. On the first project he scored 86.57 points. On the second project he scored 95.36 points. On the next one, he scored 64.11 points. How many more points must he score to go on the trip?

____

4. James was using so much water playing in the sprinkler that his mom set a limit of 15 gallons per week. On the first day he used 2.65 gallons. On the second day he used 3.21 gallons. On the third day James used 4.81 gallons. Then on the fourth day he used 1.92 gallons. How much water does James have left this week?

____

Day 37

Adding and Subtracting Money

Add or subtract the amounts below.

1.

  $103.32
+ $49.08

2.

  $315.15
+ $12.99

3.

  $296.12
+ $91.64

4.

  $454.12
  $375.36
+ $59.99

5.

  $708.51
  $623.78
+ $11.63

6.

  $551.40
  $186.17
+ $34.33

7.

  $8.84
- $3.53

8.

  $6.85
- $5.06

9.

  $9.98
- $6.16

10.

  $135.07
- $75.98

11.

  $209.47
- $105.12

12.

  $466.17
- $389.55

Day 38

Multiplying Decimals

Multiply the decimal values below. Be sure to put the decimal point in the correct position in the final answer.

1.

		8	.3
×		6	.4
	3	3	2
+ 4	9	8	0
5	3	.1	2

2.

	3	.3
×	2	.9
+		

3.

	6	.7
×	4	.4
+		

4.

	5	.9
×	3	.6
+		

5.

	4	.2	9
×		5	.3
+			

6.

	6	.8	5
×	0	.8	6
+			

7.

	3	.4	2
×		1	.3
+			

8.

	5	.3	7
×		1	.3
+			

9.

	7	.3	2
×		2	.6
+			

Chapter 6: Geometry

The most common three-dimensional figures include cones, cylinders, spheres, triangular and rectangular prisms, and cubes. When identifying and classifying three-dimensional figures it is important to be aware of the number of faces, edges, and vertices. **Faces** are the flat surfaces, **edges** are where two faces come together, and **vertices** are like corners, where more than two faces come together. For example, a rectangular prism has 6 faces, 12 edges, and 8 vertices.

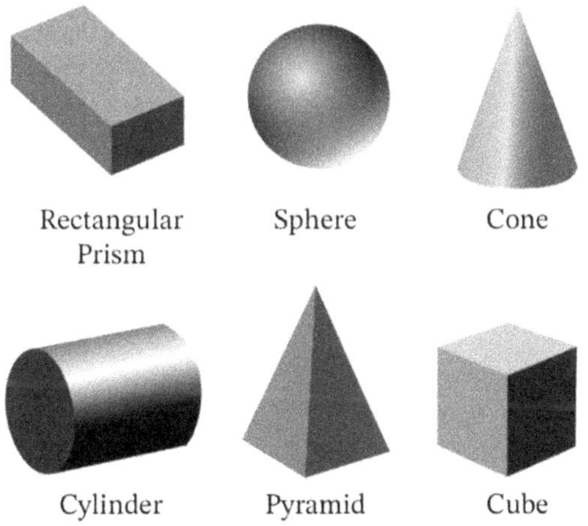

Perimeter is the distance around the outside of a two-dimensional shape. Perimeter can be calculated by finding the **sum** of a shape's **side lengths**.

The **area formula** for a **rectangle** is Area = Length × Width. This means the area can be determined by multiplying the number of **rows** by the number of **columns**. Think about the area of a rectangle as an array. For example, a rectangle that is 6 ft by 4 ft is essentially 6 columns and 4 rows. 6 groups of 4 is the same thing as 4 groups of 6, so the area can be calculated as 6×4 or 4×6. $6 \times 4 = 24$ and $4 \times 6 = 24$. The area is 24 square units.

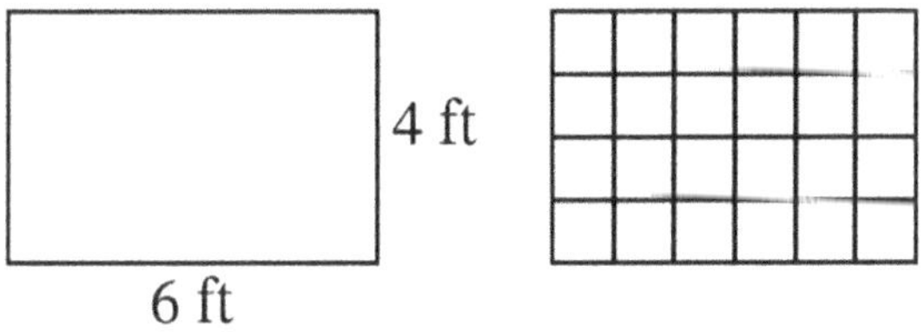

A **point** is an exact location with no size. This often appears as a dot, and is usually labeled with a letter, such as Point *A*.

A **line** is a straight path that extends in opposite directions. A line does not have a width, and it extends indefinitely in both directions. Lines are labeled with letters representing points on the line, such as $\overleftrightarrow{AB}$.

$\overleftrightarrow{AB}$

A **line segment** is a part of a line. A line segment has endpoints, beginning and ending at a point labeled with a letter, such as $\overline{AB}$.

$\overline{AB}$

A **ray** is a part of a line that starts at an endpoint and then extends out indefinitely. Rays are labeled starting with a letter at the endpoint and ending with a letter representing another point on the ray, such as $\overrightarrow{AB}$.

$\overrightarrow{AB}$

A **vertex** is formed when two straight lines or rays meet at a common endpoint.

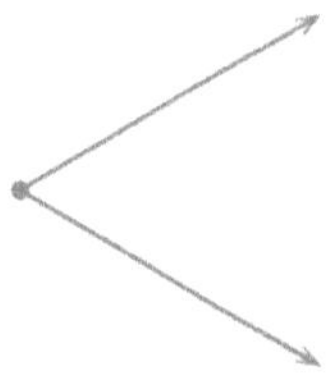

An **angle** is formed when two straight lines or rays meet at a common endpoint. The two rays below form an angle. Angles can be named by one letter or three letters. For example, the angle below is named either $\angle ABC$ or $\angle B$.

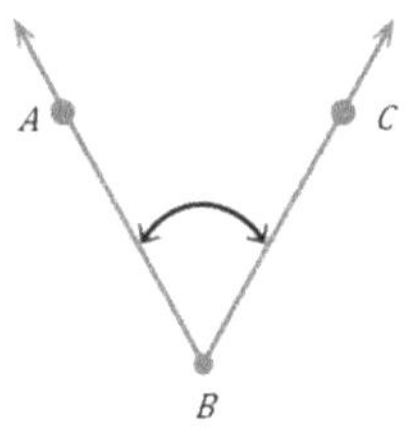

Complete one worksheet each day in this chapter. Each day should take no longer than 15 minutes to complete.

DAY 39

FINDING AREA OF SQUARES AND RECTANGLES

Find the area of each shape. Write the whole problem out.

1.

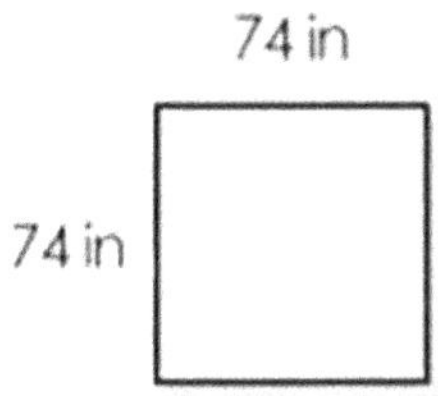

2.

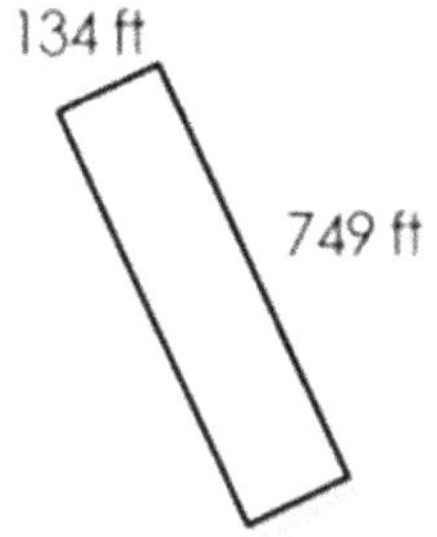

3.

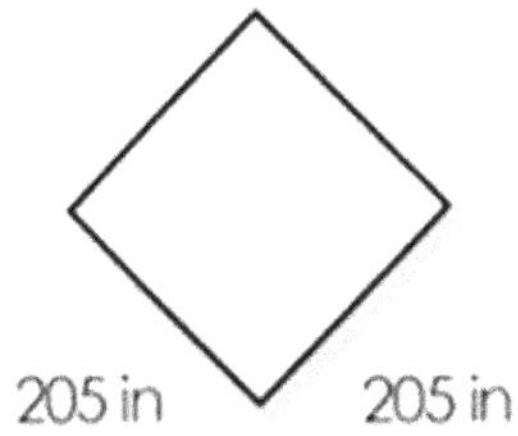

$74 \text{ in} \times 74 \text{ in} = 5{,}476 \text{ in}^2$

4.

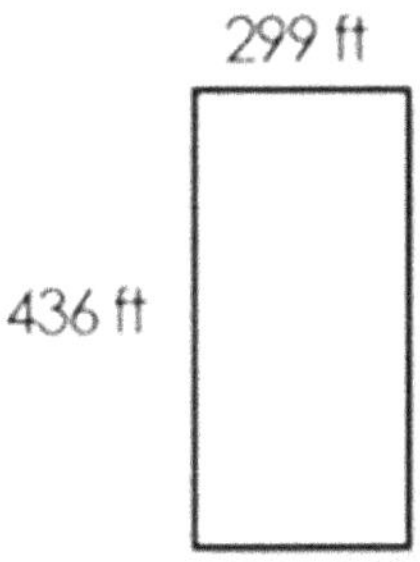

5.

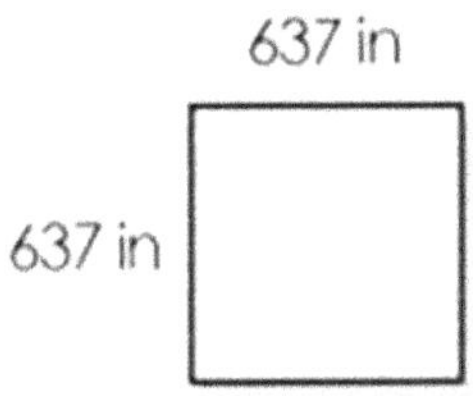

6.

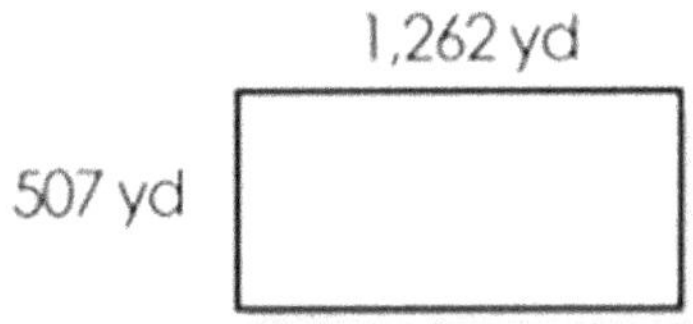

DAY 40

FINDING PERIMETER

Find the perimeter of each shape. Write the whole problem out.

1.

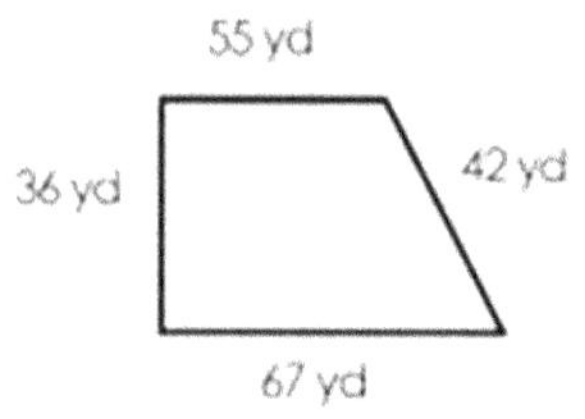

$36 + 67 + 42 + 55 = 197\text{yd}$

2.

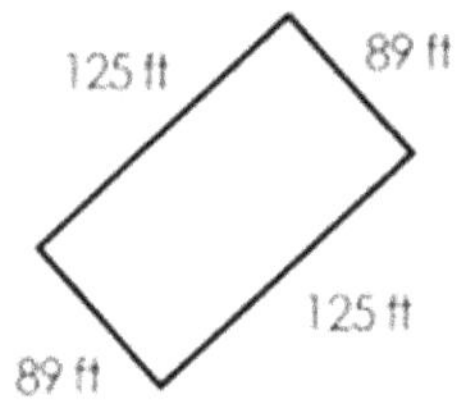

3.

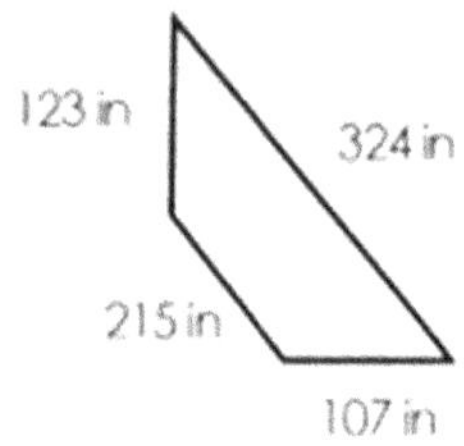

4.

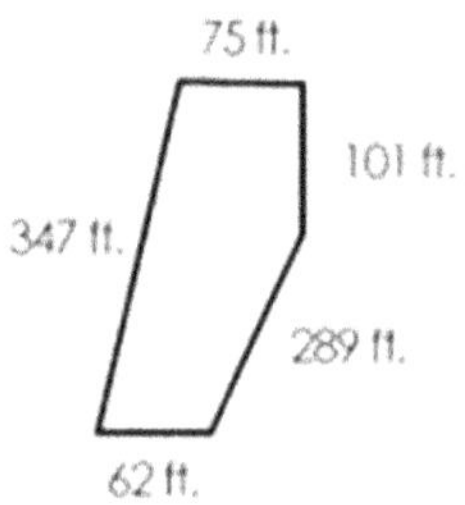

5.

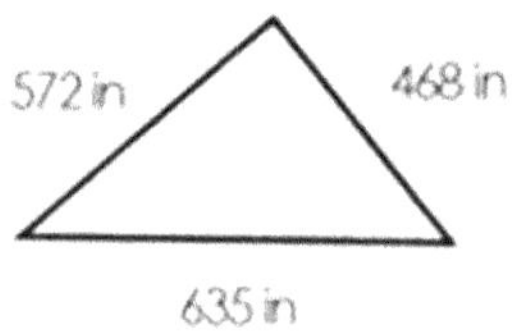

6.

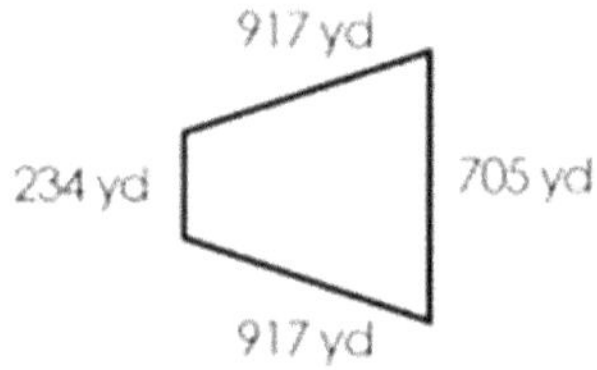

Day 41

Identifying Points, Lines and Rays

Identify if each set has a point, line, line segment, ray, or vertex. Write how many of each.

1. 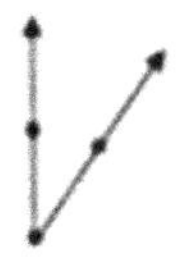

Vertex, 1

Ray, 2

Point, 3

2. 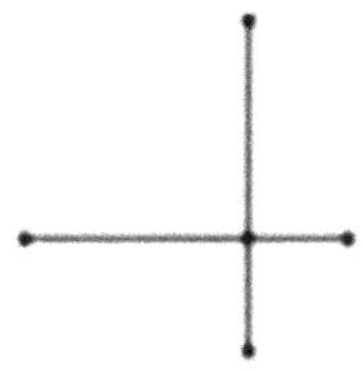

3. 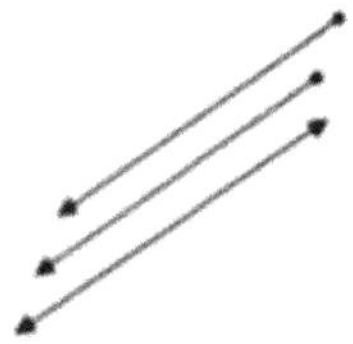

4. 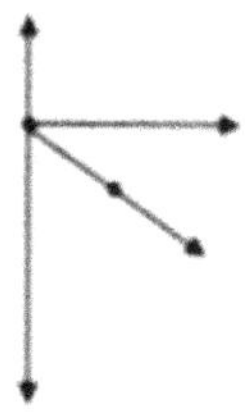

5. 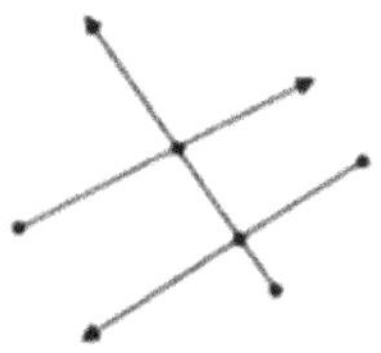

6. 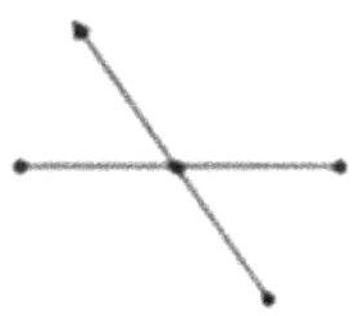

7. 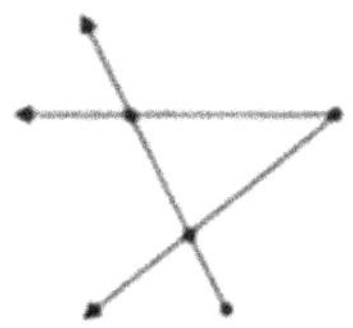

8. 

Day 42

Identifying Parts of an Angle

Identify the vertex, rays, and angle.

1.

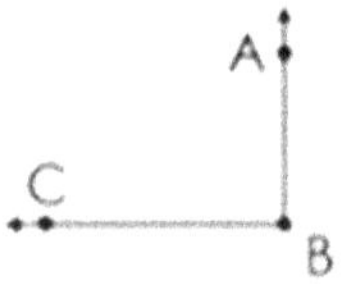

Angle: ABC

Vertex: B

Rays: AB, CB

2.

Angle: ____________

Vertex: ____________

Rays: ____________

3.

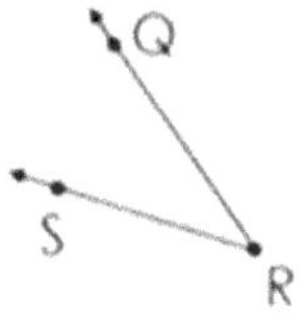

Angle: ____________

Vertex: ____________

Rays: ____________

4.

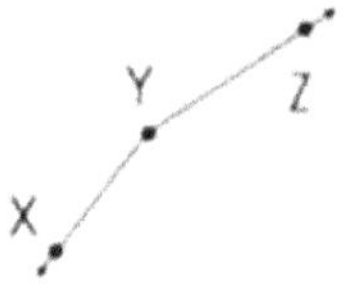

Angle: ____________

Vertex: ____________

Rays: ____________

5.

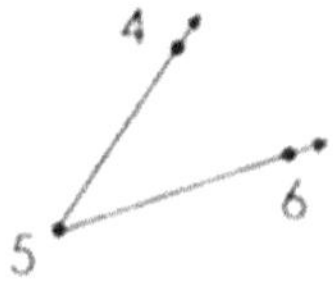

Angle: ____________

Vertex: ____________

Rays: ____________

6.

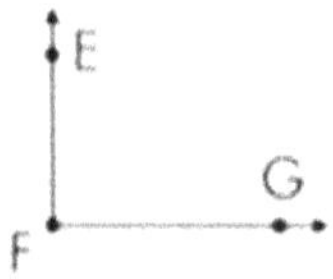

Angle: ____________

Vertex: ____________

Rays: ____________

7.

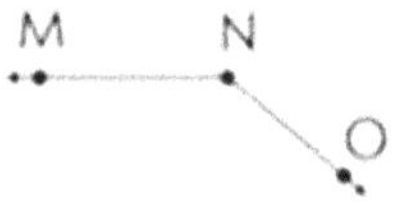

Angle: ____________

Vertex: ____________

Rays: ____________

8.

Angle: ____________

Vertex: ____________

Rays: ____________

9.

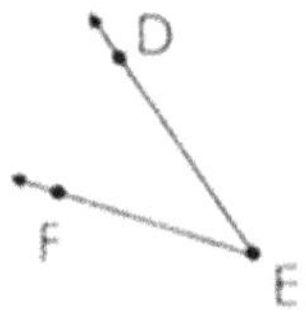

Angle: ____________

Vertex: ____________

Rays: ____________

10.

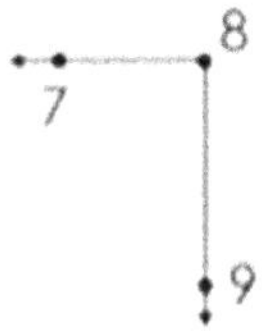

Angle: ____________

Vertex: ____________

Rays: ____________

11.

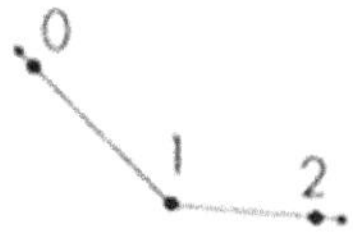

Angle: ____________

Vertex: ____________

Rays: ____________

12.

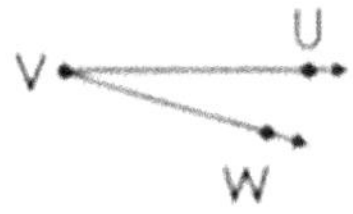

Angle: ____________

Vertex: ____________

Rays: ____________

Day 43

Identifying Solid Shapes

Write the name under each shape and then answer the questions next to it.

1.

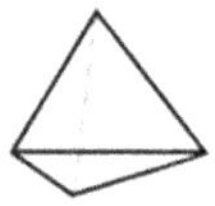

____

a. How many vertices does this shape have?

____

b. How many edges does this shape have?

____

c. How many faces does this shape have?

____

2.

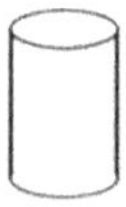

____

a. How many vertices does this shape have?

____

b. How many edges does this shape have?

____

c. How many faces does this shape have?

____

3.

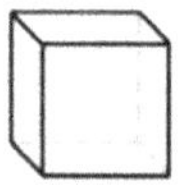

____

a. How many vertices does this shape have?

____

b. How many edges does this shape have?

____

c. How many faces does this shape have?

____

4.

____

a. How many vertices does this shape have?

____

b. How many edges does this shape have?

____

c. How many faces does this shape have?

____

5.

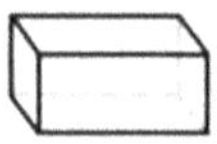

____

a. How many vertices does this shape have?

____

b. How many edges does this shape have?

____

c. How many faces does this shape have?

____

6.

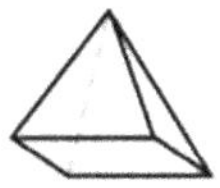

a. How many vertices does this shape have?

____

b. How many edges does this shape have?

____

c. How many faces does this shape have?

____

Chapter 7: Graphs

Data refers to a collection of information. Data can be summarized using different graphical displays such as **bar graphs, line graphs,** and **pie charts**.

Bar graphs use horizontal or vertical bars to represent the total amount for each category. For example, this bar graph shows the number of votes for a variety of snacks. The length of the bar represents the number of votes. Notice how the red popcorn bar represents 5 votes, the green crackers bar represents 2 votes, and so on. The interval should always be scaled and consistent when displaying data graphically. For example, the number of votes on this bar graph increases by 2 consistently.

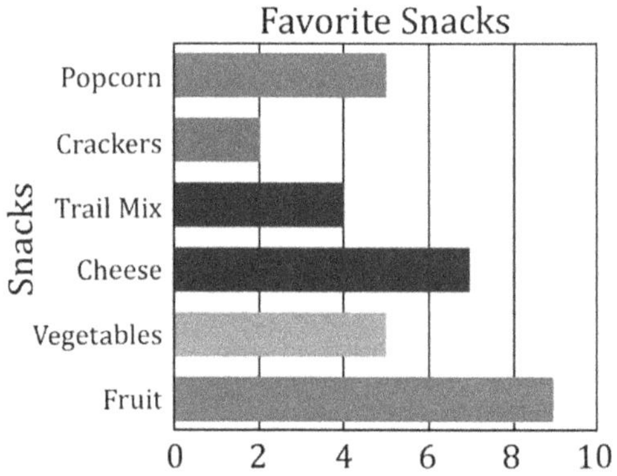

Line graphs have one or more lines of varying styles (solid or broken) to show the different values for a set of data. The individual plotted points are joined by line segments to show whether the value of the data is increasing, decreasing, or staying the same. Multiple sets of data can be graphed on the same line graph to give an easy visual comparison. An example of this would be graphing achievement test scores for different groups of students over the same time period to see which group had the greatest increase or decrease in performance from year to year (as shown below).

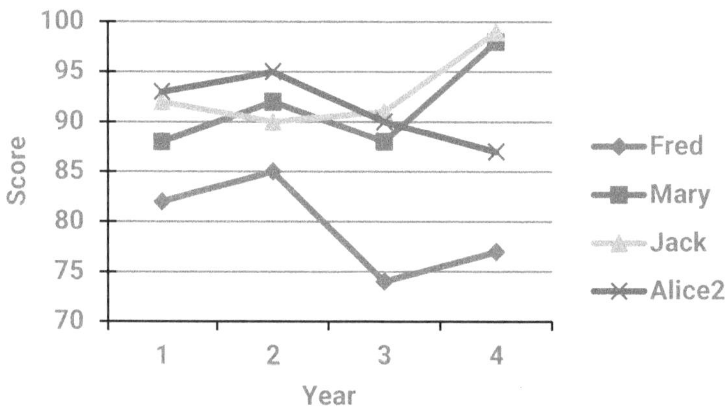

A **pie chart**, also known as a circle graph, is a type of graph that uses a circle to represent the total data and each portion of the circle is divided based on the way the data is distributed. Pie charts are commonly used with percentages and to illustrate how something is distributed across groups.

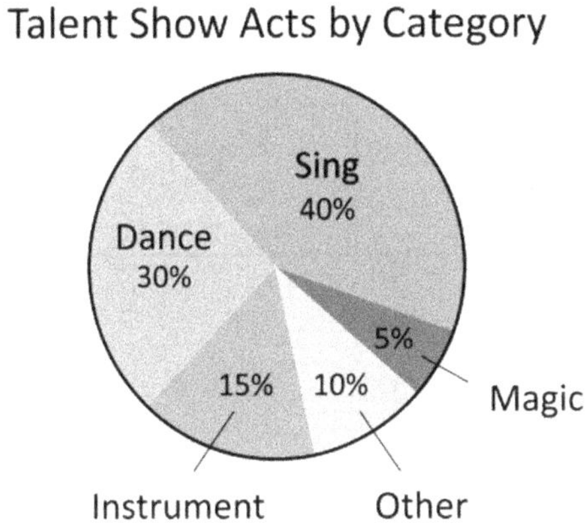

This pie chart shows the data collected from a middle school talent show. At the talent show, there were 80 students participating. 40% of the students performed a singing act, 30% of the students performed a dancing act, 15% of students played a musical instrument, 5% of students performed a magic act, and the remainder of the students, 10%, performed some other type of act.

A complete circle of the pie chart represents all students who participated in the talent show, or 100%. The size of each "slice" of the circle is proportional to the number of students who participated in a specific type of act.

The **coordinate plane** is a two-dimensional plane formed from the intersection of a vertical and horizontal line. The horizontal line is referred to as the ***x*-axis**, and the vertical line is referred to as the ***y*-axis**. The x-axis and y-axis are perpendicular to each other. The x-axis and y-axis extend out from the **origin**, which is located at the point (0,0).

Ordered pairs are points on the coordinate plane and consist of an x-coordinate and a y-coordinate. The x-coordinate indicates horizontal movement from the origin, to the right or left. The y-coordinate indicates vertical movement from the origin, up or down. In other words, the x- and y-coordinates reflect how far over and how far up an ordered pair is on the coordinate plane. For example, the point (3,2) has an x-coordinate of 3 and a y-coordinate of 2. The point (3,2) is plotted by moving 3 units to the right and 2 units up from the origin (0,0).

All ordered pairs take the form (x, y), where x describes the horizontal movement, and y describes the vertical movement. For example, consider the graph below. Start at the origin (0,0). The point (3,5) indicates a movement of 3 units horizontally to the right, and 5 units vertically up. The point (3,5) can be plotted on the coordinate plane. It is important not to swap the x and y locations within an ordered pair. For example, the location of (3,5) is not the same as the location of (5,3).

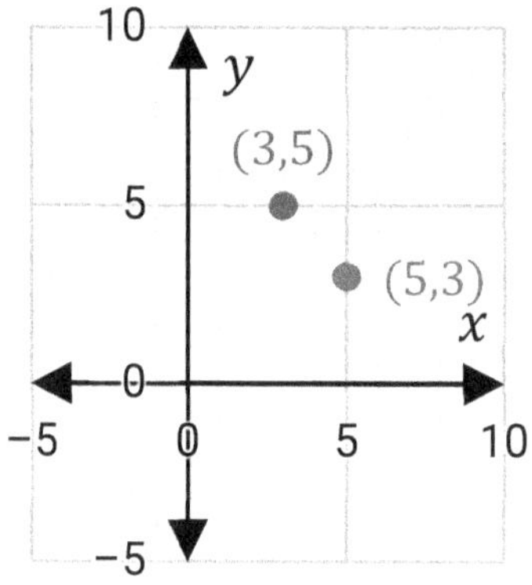

Complete one worksheet each day in this chapter. Each day should take no longer than 15 minutes to complete.

Day 44

Reading a Bar Graph

Our soccer team did well this year. Here is a graph of the points we scored each game.

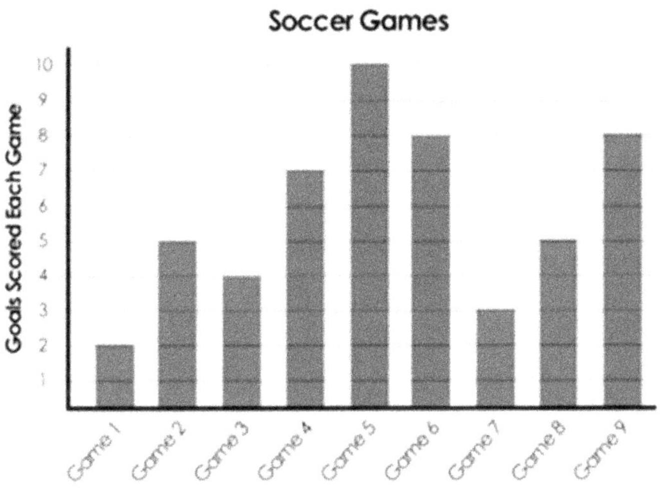

Use the bar graph above to answer the questions below.

1. In which game did we score the highest number of goals? ____________________

2. In which game did we score the lowest number of goals? ____________________

3. How many goals did we score this entire season? ____________________

4. In game 9 we scored how many more goals than in game 3?

____

5. In game 5 we scored how many more goals than in game 7?

____

6. In games 2, 3, and 4 we scored how many goals in total? ____________________

7. In games 4, 5, and 6 we scored how many goals in total? ____________________

8. In which games did we score the same number of goals? ____________________

DAY 45

CREATING A BAR GRAPH

Fundraising Totals

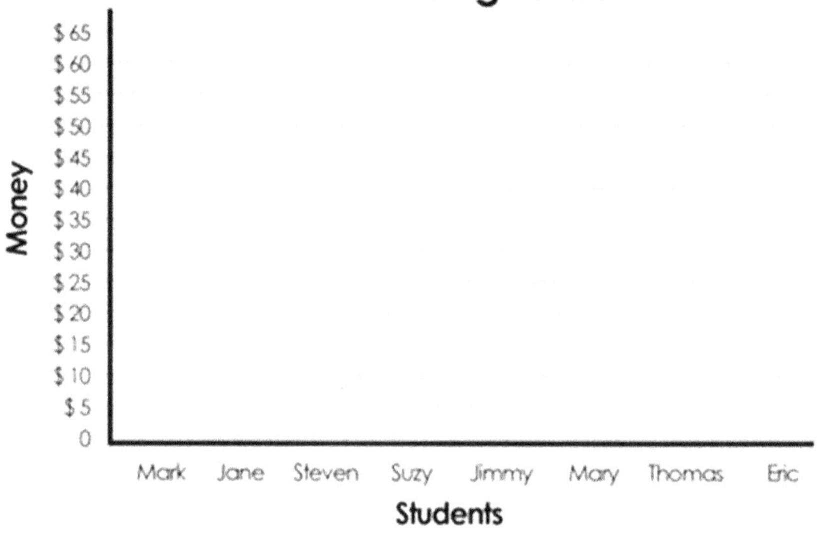

Money raised by each student	
Students	**Money**
Mark	$50
Jane	$15
Steven	$60
Suzy	$20
Jimmy	$35
Mary	$65
Thomas	$45
Eric	$5

Use the table above to make a bar graph. Then answer the questions below.

1. How much money did Thomas raise?
$__________

2. How much money did Suzy raise?
$__________

3. How much money did Jane raise?
$__________

4. How much money did Jimmy raise?
$__________

5. How much money did Mark raise?
$__________

6. How much money did Steven raise?
$__________

7. How much money did Mary raise?
$__________

8. How much money did Eric raise?
$__________

9. How much more money did Steven raise than Eric?
$__________

10. How much more money does Suzy need to raise to catch up with Jimmy?
$__________

11. How much more money does Mark need to raise to catch up with Mary?
$__________

12. How much more money did Steven raise than Jane?
$__________

13. How much more money does Suzy need to raise to catch up with Mary?
$__________

14. How much more money did Thomas raise than Jane?
$__________

Day 46

Creating a Line Graph

Numbers of People in Museum	
Time	People
9 am	20
10 am	55
11 am	30
12 pm	10
1 pm	45
2 pm	65
3 pm	60
4 pm	40
5 pm	55
6 pm	15

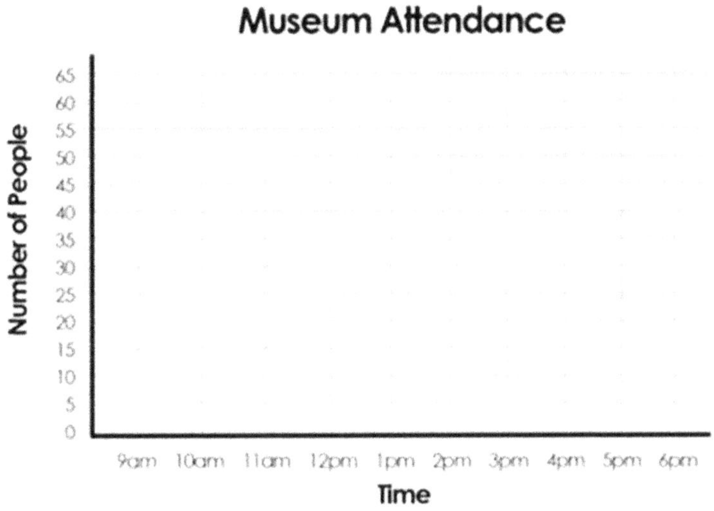

Use the table above to make a line graph. Then answer the questions below.

1. At what time was the attendance in the museum the highest?

2. At what time was the attendance in the museum the lowest?

3. How many people were in the museum at noon?

4. How many people were in the museum at 9am?

5. How many people were in the museum at 4pm?

6. How many people were in the museum at 5pm compared to 11am?

7. From 1pm to 2pm, how many people showed up at the museum?

8. From 5pm to 6pm, how many people left the museum?

9. During what one-hour time period did the attendance increase the most?

10. During what one-hour time period did the attendance decrease the most?

Day 47

Identifying Points on a Graph

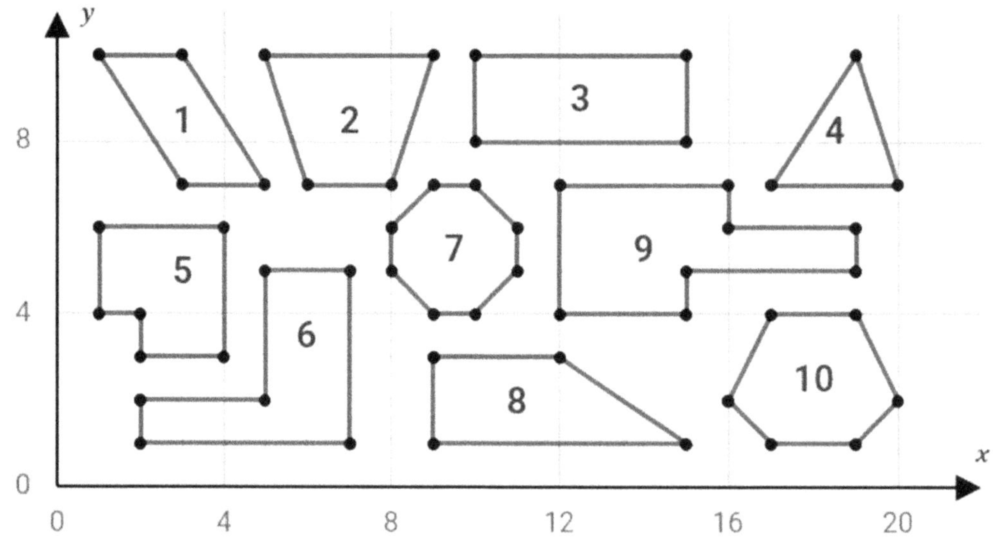

Using the graph above, write out the points that make up each shape.

1. (1,10), (3,7), (5,7), (3,10) ______
2. ______
3. ______
4. ______
5. ______
6. ______
7. ______
8. ______
9. ______
10. ______

DAY 48

PLOTTING POINTS ON A GRAPH

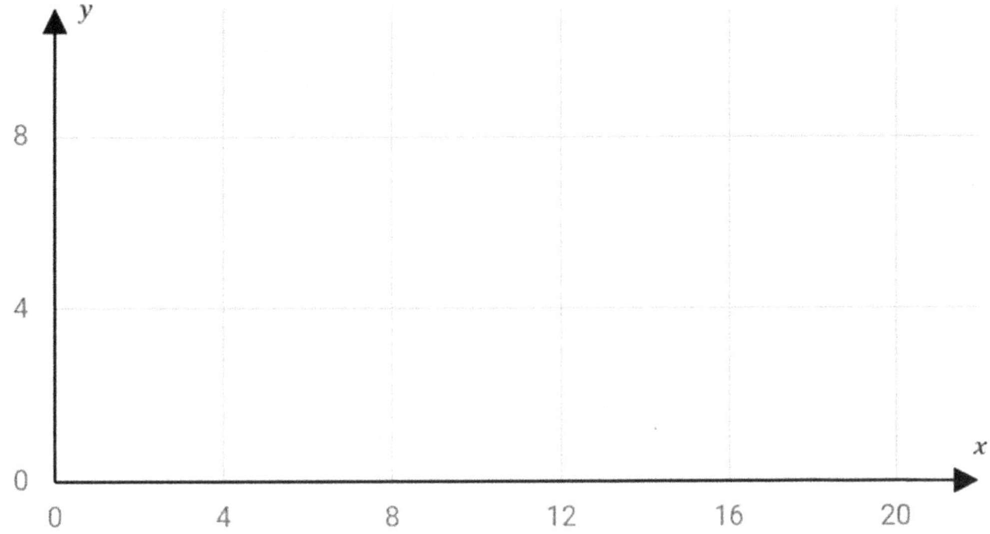

Use the coordinate below to plot the points on the graph above. Then draw a line to connect the points for each question. Then write the name of each shape that was made on the line.

1. (1,4), (2,1), (5,1), (6,4) Trapezoid ______________________

2. (7,3), (9,3), (7,1), (9,1) ______________________

3. (8,6), (12,6), (8,4), (12,4) ______________________

4. (16,7), (18,10), (20,7) ______________________

5. (6,7), (9,10), (12,10), (9,7) ______________________

6. (14,8), (13,5), (14,2), (15,5) ______________________

7. (16,4), (18,5), (20,4), (17,1), (19,1) ______________________

DAY 49

READING A PIE CHART

Use the pie graphs below to answer the questions.

STUDENTS' FAVORITE CLASSES

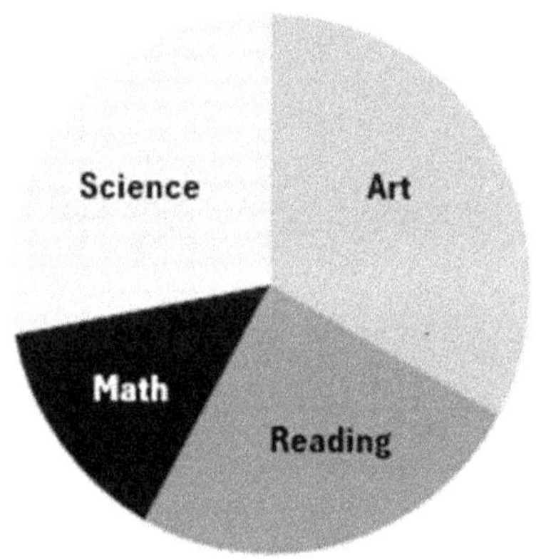

1. Which class do the students like most?

___

2. Which class do the students like least?

___

3. What is the second favorite class?

___

4. Do the students like math or reading more?

___

KIDS' FAVORITE PETS

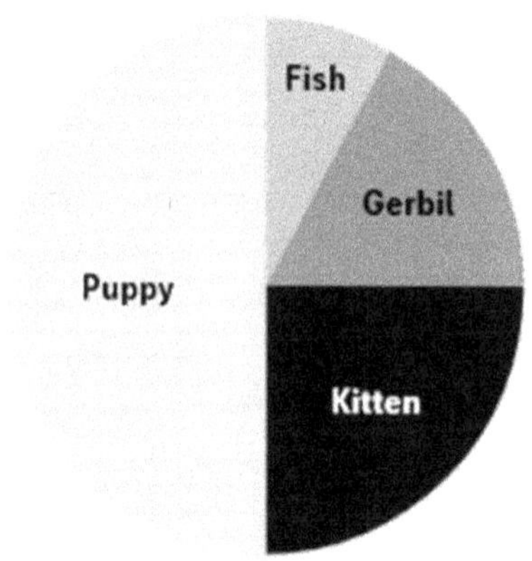

5. What is the kids' favorite pet?

___

6. What is the kids' least favorite pet?

___

7. What is the kids' second favorite pet?

___

8. Do more kids want a gerbil or a fish?

___

Day 50

Creating a Pie Chart

Use the information provided to fill each pie graph. For the top two graphs, place the corresponding number in each section of the graph. For the bottom two graphs, partition the graph into the corresponding sections and label with the corresponding numbers.

1. Mars: 50%
2. Mercury: 6%
3. Venus: 10%
4. Neptune: 34%

MOST STUDIED PLANETS

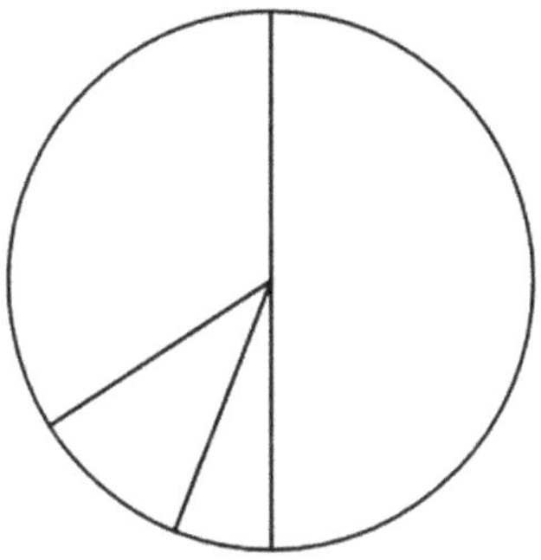

10. School Bus: 42%
11. Car: 35%
12. Bike: 15%
13. Walk: 8%

HOW STUDENTS GET TO SCHOOL

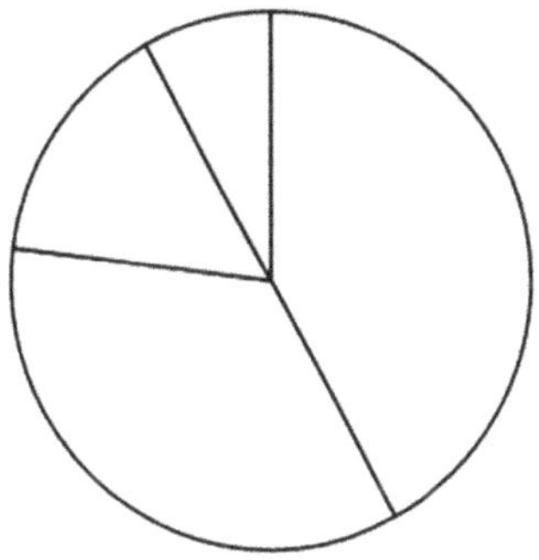

5. Comedy: 35%
6. Action: 18%
7. Drama: 8%
8. Animation: 24%
9. Science Fiction: 15%

FAVORITE MOVIE TYPES IN SCHOOLS

14. Bass: 20%
15. Catfish: 27%
16. Trout: 14%
17. Guppies: 32%
18. Turtles: 7%

FISH POPULATION IN THE POND

Workbook Answers

Chapter 1: Whole Numbers

Day 1

Place Value 1-100,000,000

1.	7	4	8	4	2	2	7	1	9
	Hundred Millions	Ten Millions	Millions	Hundred Thousands	Ten Thousands	Thousands	Hundreds	Tens	Ones
2.	3	2	9	6	0	8	1	1	4
	Hundred Millions	Ten Millions	Millions	Hundred Thousands	Ten Thousands	Thousands	Hundreds	Tens	Ones
3.	1	2	4	3	7	5	2	7	7
	Hundred Millions	Ten Millions	Millions	Hundred Thousands	Ten Thousands	Thousands	Hundreds	Tens	Ones
4.	7	4	1	5	8	8	3	7	9
	Hundred Millions	Ten Millions	Millions	Hundred Thousands	Ten Thousands	Thousands	Hundreds	Tens	Ones
5.	5	0	4	2	6	7	3	3	2
	Hundred Millions	Ten Millions	Millions	Hundred Thousands	Ten Thousands	Thousands	Hundreds	Tens	Ones

Day 2

Rounding Up to 100,000

1. 3,000
2. 9,000
3. 1,000
4. 8,000
5. 3,000
6. 4,000
7. 6,000
8. 8,000
9. 6,000
10. 2,000

11. 40,000
12. 20,000
13. 90,000
14. 60,000
15. 70,000
16. 80,000
17. 30,000
18. 40,000
19. 20,000
20. 50,000

21. 300,000
22. 600,000
23. 200,000
24. 500,000
25. 200,000
26. 300,000
27. 900,000
28. 600,000
29. 800,000
30. 100,000

Day 3

Estimating

1. 900 + 400 = 1,300
2. 500 + 500 = 1,000
3. 100 + 600 = 700
4. 400 + 700 = 1,100
5. 900 – 500 = 400
6. 800 – 200 = 600
7. 600 – 500 = 100
8. 400 – 200 = 200

Day 4

5-Digit Addition

1. 104,654
2. 188,944
3. 144,072
4. 189,476
5. 104,323
6. 103,894
7. 156,084
8. 140,874
9. 184,913
10. 147,781
11. 57,012
12. 183,425
13. 131,969
14. 101,039
15. 184,161
16. 115,576
17. 83,250
18. 68,871
19. 174,540
20. 264,261

Day 5

6-Digit Addition

1. 596,052
2. 977,698
3. 1,160,308
4. 1,097,102
5. 1,089,063
6. 1,283,759
7. 1,336,765
8. 845,509
9. 1,051,175

Day 6

5-Digit Subtraction

1. 2,786
2. 25,038
3. 21,376
4. 36,508
5. 20,756
6. 16,804
7. 20,507
8. 4,557
9. 37,509
10. 38,809
11. 18,499
12. 26,055

Day 7

6-Digit Subtraction

1. 131,482
2. 507,851
3. 277,758
4. 34,013
5. 778,375
6. 191,436
7. 411,246
8. 352,113
9. 221,963

Chapter 2: Multiplication

Day 8

Multiplication

1. 260
2. 147
3. 108
4. 552
5. 456
6. 392
7. 116
8. 160
9. 476
10. 747

Day 9

Multiplying 2-Digit Numbers by 2-Digit Numbers

1. 735
2. 1,536
3. 1,944
4. 3,216
5. 5,525
6. 3,404
7. 3,354
8. 1,015
9. 3,196
10. 1,800
11. 2,916
12. 5,568
13. 2,666
14. 8,740
15. 1,638

Day 10

Multiplication Word Problems

1. 42
2. 640
3. 57
4. 72

Day 11

Multiplying 3-Digit Numbers by 2-Digit Numbers

1 7,383
2 4,921
3 22,016
4 16,524
5 15,200
6 24,999
7 33,418
8 5,640
9 21,204
10 53,952
11 74,036
12 37,905
13 23,970
14 50,007
15 47,288

Day 12

Multiplying 4-Digit Numbers by 2-Digit Numbers

1. 77,532
2. 102,915
3. 167,684
4. 253,425
5. 90,364
6. 376,467
7. 296,472
8. 230,278
9. 156,581
10. 254,910
11. 238,908
12. 404,247

Day 13

Multiplying 3-Digit Numbers by 3-Digit Numbers

1. 157,680
2. 222,221
3. 79,530
4. 269,346
5. 63,784
6. 120,825
7. 568,874
8. 182,400
9. 145,148

Day 14

Multiplying 4-Digit Numbers by 3-Digit Numbers

1. 418,284
2. 250,355
3. 584,168
4. 826,428
5. 1,660,480
6. 1,604,106
7. 785,436
8. 588,707
9. 1,668,172

Chapter 3: Division

Day 15

Division with a 2-Digit Dividend

1. 13 r 2
2. 12 r 3
3. 17 r 2
4. 10 r 2
5. 39
6. 11 r 3
7. 14 r 3
8. 21 r 1
9. 15 r 1
10. 41 r 1
11. 10 r 7
12. 12 r 5

Day 16

Division Word Problems

1. 8 trips
2. 5 days
3. 7 apples
4. 9 cookies

Day 17

Dividing 3-Digit Dividends by a 1-Digit Divisor

1. 153 r 2
2. 32
3. 196
4. 139 r 2
5. 90 r 5
6. 148 r 4
7. 188
8. 148 r 1
9. 207 r 1
10. 232 r 2
11. 111 r 1
12. 140 r 6
13. 423
14. 44 r 2
15. 204

Day 18

Dividing 4-Digit Dividends by a 1-Digit Divisor

1	1,241 r 2	5	1,325 r 3	9	450 r 4
2	569 r 5	6	2,178 r 3	10	1,867 r 1
3	1,522 r 1	7	632	11	1,142 r 2
4	630 r 1	8	968 r 1	12	999 r 3

Day 19

Division with 2-Digit Divisors

1.	21 r 8
2.	19 r 10
3.	20 r 9
4.	37 r 5
5.	15 r 24
6.	13 r 24
7.	16 r 16
8.	12 r 9
9.	10 r 53
10.	19 r 19

Day 20

Dividing 3-Digit Dividends by a 2-Digit Divisor

1.	22 r 5	5.	43 r 1	9.	21 r 20
2.	24 r 18	6.	22 r 7	10.	17 r 8
3.	20 r 9	7.	22 r 4	11.	10 r 22
4.	5 r 4	8.	22 r 9	12.	10 r 26

Day 21

Dividing 4-Digit Dividends by a 2-Digit Divisor

1.	207	5.	226 r 12	9.	203 r 30
2.	148 r 6	6.	259 r 11	10.	118 r 3
3.	137 r 40	7.	213 r 14	11.	161 r 17
4.	111 r 27	8.	291 r 3	12.	205 r 10

Chapter 4: Fractions

Day 22

Adding and Subtracting Fractions with Common Denominators

1. $\frac{13}{15}$
2. $\frac{23}{37}$
3. $\frac{49}{80}$
4. $\frac{100}{109}$
5. $\frac{30}{64}$
6. $\frac{14}{21}$
7. $\frac{283}{759}$
8. $\frac{26}{56}$
9. $\frac{79}{207}$
10. $\frac{13}{49}$
11. $\frac{167}{277}$
12. $\frac{238}{534}$

Day 23

Fraction Word Problems

1. Tomatoes: $\frac{13}{28}$ Eggplants: $\frac{6}{28}$ Potatoes: $\frac{9}{28}$
2. Blue Jays: $\frac{3}{15}$ Sparrows: $\frac{9}{15}$ Ducks: $\frac{6}{15}$ Eagles: $\frac{1}{15}$
3. Bass: $\frac{2}{17}$ Trout: $\frac{6}{17}$ Guppies: $\frac{5}{17}$ Goldfish: $\frac{4}{17}$
4. Hamburgers: $\frac{15}{76}$ Chicken Wings: $\frac{22}{76}$ Sausages: $\frac{39}{76}$

Day 24

Equivalent Fractions

1. 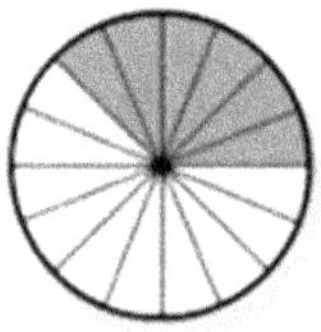

2. 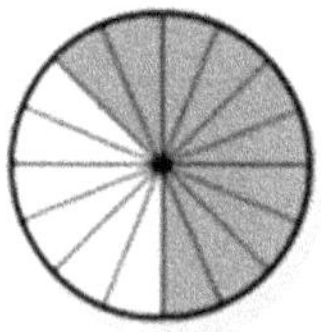

3. $\frac{9}{9}, \frac{2}{3}$
4. $\frac{4}{4}, \frac{1}{3}$
5. $\frac{5}{5}, \frac{4}{7}$
6. $\frac{3}{3}, \frac{3}{8}$
7. $\frac{3}{3}, \frac{5}{7}$
8. $\frac{8}{8}, \frac{2}{5}$
9. $\frac{3}{5}, \frac{4}{5}$
10. $\frac{3}{7}, \frac{5}{7}$
11. $\frac{3}{5}, \frac{2}{5}$
12. $\frac{5}{7}, \frac{2}{7}$

Day 25

Adding and Subtracting Fractions with Different Denominators – Using Division

1. $\frac{5}{7} + \frac{2}{7} = \frac{7}{7}$
2. $\frac{1}{5} + \frac{2}{5} = \frac{3}{5}$
3. $\frac{7}{10} + \frac{2}{10} = \frac{9}{10}$
4. $\frac{10}{12} + \frac{2}{12} = \frac{12}{12}$
5. $\frac{2}{9} + \frac{3}{9} = \frac{5}{9}$
6. $\frac{5}{7} + \frac{2}{7} = \frac{7}{7}$
7. $\frac{9}{9} - \frac{6}{9} = \frac{3}{9}$
8. $\frac{8}{9} - \frac{3}{9} = \frac{5}{9}$
9. $\frac{6}{7} - \frac{3}{7} = \frac{3}{7}$
10. $\frac{8}{9} - \frac{2}{9} = \frac{6}{9}$
11. $\frac{5}{6} - \frac{2}{6} = \frac{3}{6}$
12. $\frac{8}{9} - \frac{5}{9} = \frac{3}{9}$

Day 26

Finding Common Denominators using Multiplication

1. 12
2. 12
3. 10
4. 40
5. 5
6. 90
7. 80
8. 16
9. 28
10. 90
11. 14
12. 55
13. 128
14. 15
15. 50
16. 80
17. 12
18. 15
19. 6
20. 16

Day 27

Adding and Subtracting Fractions with Different Denominators – Using Multiplication

1. $\frac{4}{8}+\frac{1}{8}=\frac{5}{8}$
2. $\frac{4}{12}+\frac{6}{12}=\frac{10}{12}$
3. $\frac{6}{21}+\frac{5}{21}=\frac{11}{21}$
4. $\frac{4}{28}+\frac{21}{28}=\frac{25}{28}$
5. $\frac{30}{50}+\frac{12}{50}=\frac{42}{50}$
6. $\frac{6}{24}+\frac{15}{24}=\frac{21}{24}$
7. $\frac{36}{49}-\frac{21}{49}=\frac{15}{49}$
8. $\frac{45}{54}-\frac{29}{54}=\frac{16}{54}$
9. $\frac{63}{81}-\frac{54}{81}=\frac{9}{81}$
10. $\frac{54}{63}-\frac{35}{63}=\frac{19}{63}$
11. $\frac{27}{48}-\frac{24}{48}=\frac{3}{48}$
12. $\frac{70}{100}-\frac{50}{100}=\frac{20}{100}$

Day 28

Mixed Numbers with Common Denominators

1. $5\frac{10}{12}$
2. $6\frac{4}{9}$
3. $7\frac{4}{5}$
4. $2\frac{20}{23}$
5. $13\frac{6}{8}$
6. $13\frac{10}{15}$
7. $13\frac{30}{39}$
8. $12\frac{9}{10}$
9. $4\frac{5}{19}$
10. $3\frac{5}{8}$
11. $3\frac{6}{16}$
12. $5\frac{4}{6}$
13. $5\frac{4}{25}$
14. $11\frac{1}{10}$
15. $8\frac{27}{44}$

Day 29

Reducing Mixed Numbers with Common Denominators

1. $5\frac{5}{10} = 5\frac{1}{2}$

2. $9\frac{9}{12} = 9\frac{3}{4}$

3. $9\frac{6}{18} = 9\frac{1}{3}$

4. $8\frac{4}{8} = 8\frac{1}{2}$

5. $12\frac{12}{24} = 12\frac{1}{2}$

6. $15\frac{18}{36} = 15\frac{1}{2}$

7. $3\frac{8}{16} = 3\frac{1}{2}$

8. $12\frac{4}{12} = 12\frac{1}{3}$

9. $8\frac{14}{28} = 8\frac{1}{2}$

10. $12\frac{6}{9} = 12\frac{2}{3}$

11. $39\frac{28}{42} = 39\frac{2}{3}$

12. $18\frac{18}{63} = 18\frac{2}{7}$

Day 30

Converting Mixed Numbers and Improper Fractions

1.	$\frac{13}{3}$	9.	7	17.	$2\frac{3}{6}$
2.	$\frac{38}{4}$	10.	4	18.	$2\frac{2}{4}$
3.	$\frac{13}{2}$	11.	3	19.	$3\frac{1}{2}$
4.	$\frac{32}{5}$	12.	9	20.	$2\frac{3}{8}$
5.	$\frac{95}{10}$	13.	8	21.	$5\frac{3}{4}$
6.	$\frac{21}{5}$	14.	31	22.	$8\frac{2}{3}$
7.	$\frac{25}{7}$	15.	5	23.	$5\frac{5}{9}$
8.	$\frac{14}{3}$	16.	42	24.	$5\frac{6}{8}$

Chapter 5: Decimals

Day 31

Arranging and Adding Decimals

1. 19.44
2. 17.82
3. 18.61
4. 22.83
5. 264.57
6. 85.62
7. 93.31
8. 115.88
9. 1,000.77

Day 32

Decimal Addition Word Problems

1. points: 75.4 rebounds: 16.2
2. 18.5
3. 19.55
4. 40.34

Day 33

Adding Decimals

1. 165.93
2. 130.62
3. 337.26
4. 443.47
5. 898.79
6. 1,328.24
7. 913.05
8. 1,676.61
9. 98.011
10. 47.257
11. 134.537
12. 110.116

Day 34

Arranging and Subtracting Decimals

1. 49.97
2. 27.68
3. 13.89
4. 15.96
5. 1.17
6. 30.6
7. 18.17
8. 46.75
9. 17.56

Day 35

Subtracting Decimals

1. 49.33
2. 326.39
3. 105.26
4. 123.73
5. 241.84
6. 341.48
7. 44.791
8. 20.014
9. 13.269
10. 5.9
11. 21.28
12. 36.574

Day 36

Decimal Subtraction Word Problems

1. 1.35
2. 26.82
3. 53.96
4. 2.41

Day 37

Adding and Subtracting Money

1. \$152.40
2. \$328.14
3. \$387.76
4. \$889.47
5. \$1,343.92
6. \$771.90
7. \$5.31
8. \$1.79
9. \$3.82
10. \$59.09
11. \$104.35
12. \$76.62

Day 38

Multiplying Decimals

1. 53.12
2. 9.57
3. 29.48
4. 21.24
5. 22.737
6. 5.891
7. 4.446
8. 34.368
9. 19.032

Chapter 6: Geometry

Day 39

Finding Area of Squares and Rectangles

1. $74 \text{ in} \times 74 \text{ in} - 5{,}476 \text{ in}^2$
2. $134 \text{ ft} \times 749 \text{ ft} = 100{,}366 \text{ ft}^2$
3. $205 \text{ in} \times 205 \text{ in} = 42{,}025 \text{ in}^2$
4. $299 \text{ ft} \times 436 \text{ ft} = 130{,}364 \text{ ft}^2$
5. $637 \text{ in} \times 637 \text{ in} = 405{,}769 \text{ in}^2$
6. $507 \text{ yd} \times 1{,}262 \text{ yd} = 639{,}834 \text{ yd}^2$

Day 40

Finding Perimeter

1. $55 + 36 + 42 + 67 = 197$ yd
2. $125 + 89 + 89 + 125 = 428$ ft
3. $123 + 215 + 107 + 324 = 769$ in
4. $65 + 347 + 62 + 289 = 874$ ft
5. $572 + 635 + 468 = 1{,}675$ in
6. $917 + 234 + 705 + 917 = 2{,}773$ yd

Day 41

Identifying Points, Lines, and Rays

1. ray- 2, vertex- 1, point- 3
2. line segment-2, vertex-1, point -5
3. line-1, ray-2, point-2
4. line-1, ray-2, vertex-1, point-2
5. ray-3, vertex-2, point-5
6. ray-1, line segment-1, vertex-1, point-4
7. ray- 3, vertex-3, point-4
8. line-2, ray-2, point-2

Day 42

Identifying Parts of an Angle

1.	Angle: ABC	Vertex: B	Rays: AB, CB
2.	Angle: 123	Vertex: 2	Rays: 12, 23
3.	Angle: QRS	Vertex: R	Rays: SR, QR
4.	Angle: XYZ	Vertex: Y	Rays: XY, YZ
5.	Angle: 456	Vertex: 5	Rays: 45, 56
6.	Angle: EFG	Vertex: F	Rays: EF, EG
7.	Angle: MNO	Vertex: N	Rays: MN, NO
8.	Angle: 678	Vertex: 7	Rays: 67, 78
9.	Angle: DEF	Vertex: E	Rays: FE, DE
10.	Angle: 789	Vertex: 8	Rays: 78, 89
11.	Angle: 012	Vertex: 1	Rays: 01, 12
12.	Angle: UVW	Vertex: V	Rays: UV, VW

Day 43

Identifying Solid Shapes

1. Pyramid
   a. 4
   b. 6
   c. 4

2. Cylinder
   a. 0
   b. 2
   c. 3

3. Cube
   a. 8
   b. 12
   c. 6

4. Cone
   a. 1
   b. 1
   c. 2

5. Rectangular Prism
   a. 8
   b. 12
   c. 6

6. Pyramid
   a. 5
   b. 8
   c. 5

Chapter 7: Graphs

Day 44

Reading a Bar Graph

1. Game 5
2. Game 1
3. 52 Goals
4. 4 Goals
5. 7 Goals
6. 16 Goals
7. 25 Goals
8. Games 2 and 8

Day 45

Creating a Bar Graph

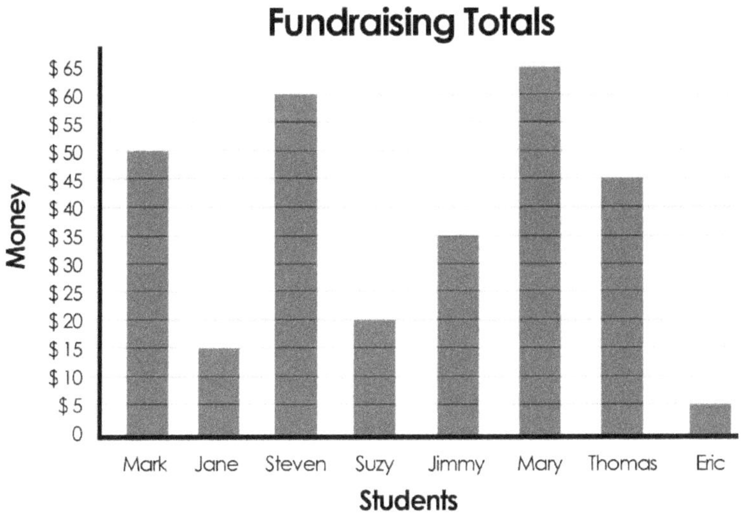

1. $45
2. $20
3. $15
4. $35
5. $50
6. $60
7. $65
8. $5
9. $55
10. $15
11. $15
12. $45
13. $45
14. $30

DAY 46

CREATING A LINE GRAPH

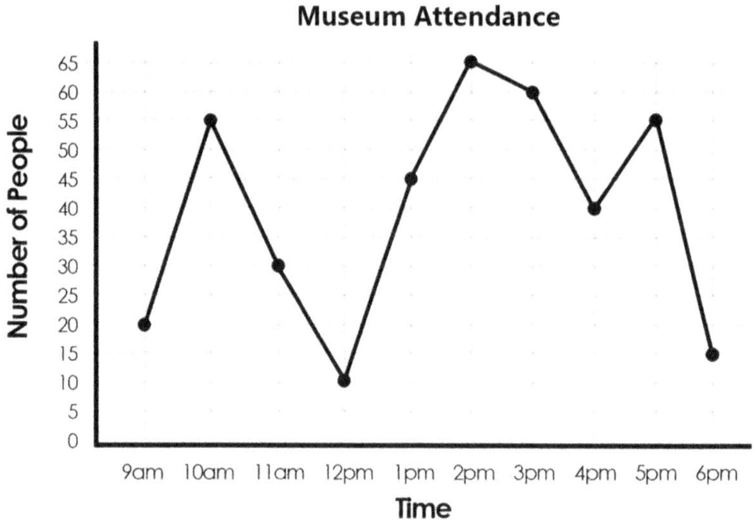

1. 2pm
2. 12pm
3. 10
4. 20
5. 40
6. 25
7. 20
8. 40
9. 12pm – 1pm
10. 5pm – 6pm

DAY 47

IDENTIFYING POINTS ON A GRAPH

1. (1,10), (3,7), (5,7), (3,10)
2. (5,10), (9,10), (6,7), (8,7)
3. (10,10), (10,8), (15,10), (15,8)
4. (17,7), (19,10), (20,7)
5. (1,6), (1,4), (2,4), (2,3), (4,3), (4,6)
6. (2,2), (2,1), (5,5), (5,2), (7,5), (7,1)
7. (8,6), (8,5), (9,7), (9,4), (10,7), (10,4), (11,6), (11,5)
8. (9,3), (9,1), (12,3), (15,1)
9. (12,7), (12,4), (15,5), (15,4), (16,7), (16,6), (19,6), (19,5)
10. (16,2), (17,4), (17,1), (19,4), (19,1), (20,2)

Day 48

Plotting Points on a Graph

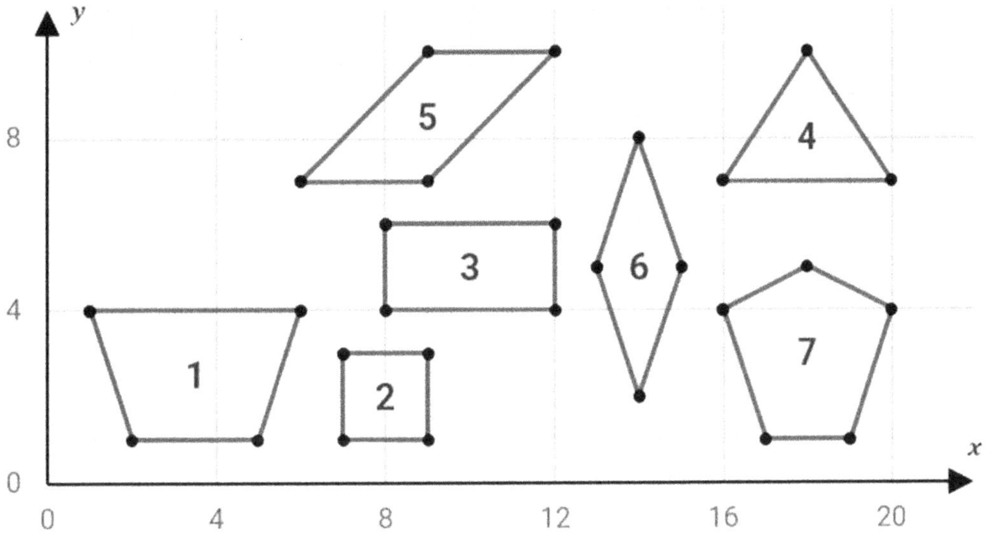

1. trapezoid
2. square
3. rectangle
4. triangle
5. parallelogram
6. rhombus
7. pentagon

Day 49

Reading a Pie Chart

1. Art
2. Math
3. Science
4. Reading
5. Puppy
6. Fish
7. Kitten
8. Gerbil

Day 50

Creating a Pie Chart

MOST STUDIED PLANETS

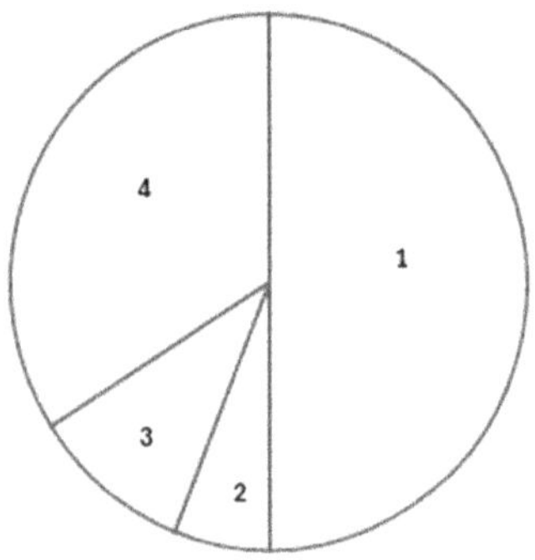

HOW STUDENTS GET TO SCHOOL

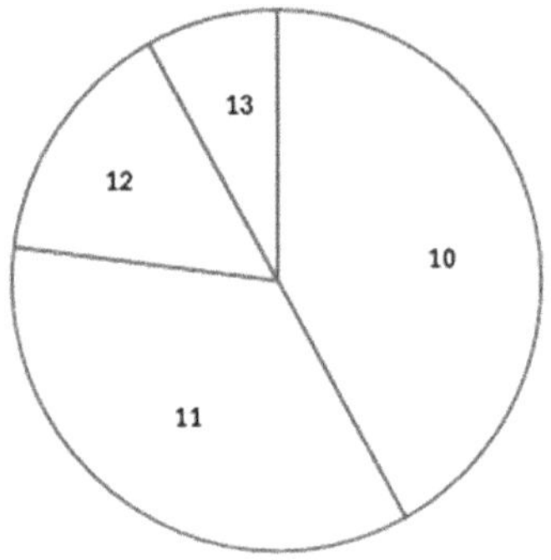

FAVORITE MOVIE TYPES IN SCHOOLS

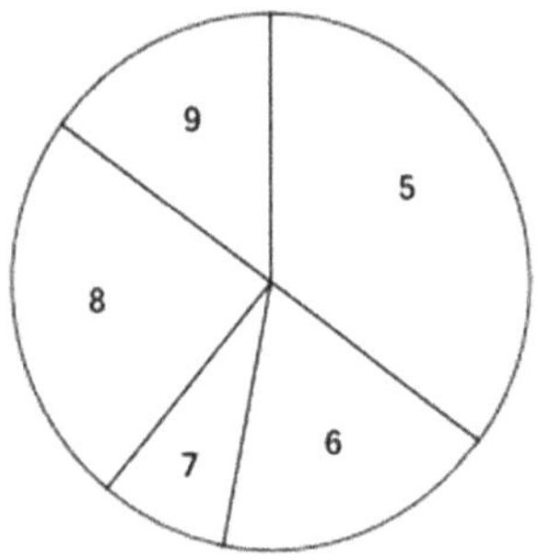

FISH POPULATION IN THE POND

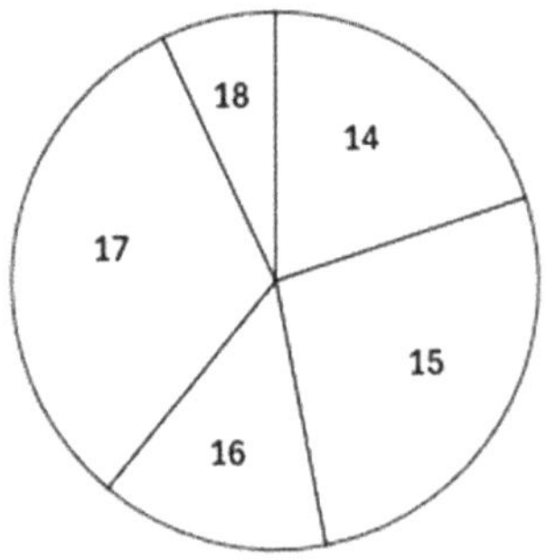

www.ingramcontent.com/pod-product-compliance
Lightning Source LLC
LaVergne TN
LVHW082357200926
848481LV00012B/3589
* 9 7 8 1 5 1 6 7 3 1 4 5 9 *